FLOWERS

MORPHOLOGY, EVOLUTIONARY DIVERSIFICATION AND IMPLICATIONS FOR THE ENVIRONMENT

BOTANICAL RESEARCH AND PRACTICES

Additional books in this series can be found on Nova's website under the Series tab.

Additional e-books in this series can be found on Nova's website under the e-book tab.

FLOWERS

MORPHOLOGY, EVOLUTIONARY DIVERSIFICATION AND IMPLICATIONS FOR THE ENVIRONMENT

TEODOR BERNTSEN

AND

KAJ ALSVIK

EDITORS

nova publishers

New York

Library of Congress Cataloging-in-Publication Data

Flowers : morphology, evolutionary diversification and implications for the environment / editors: Teodor Berntsen and Kaj Alsvik.
     p. cm.
  Includes index.
  ISBN: 978-1-62808-798-7 (hardcover)
  1. Flowers--Morphology. 2. Flowers--Evolution. 3. Angiosperms--Morphology. 4.
Angiosperms--Evolution. I. Berntsen, Teodor. II. Alsvik, Kaj.
  QK653. F585 2013
  582--dc23
                          2013030606

Published by Nova Science Publishers, Inc. † New York

CONTENTS

PREFACE

In this book, the authors present topical research in the study of the morphology, evolutionary diversification and implications for the environments of flowers. Topics discussed in this compilation include the bioactive components from asteraceae flowers; the classification, phylogenetic status and uses as ornamental groundcover of the arachis species; understanding the role of pigments in flowers; flowers as sources of therapeutic molecules; the floral development of sauvagesia (ochnaceae) revealing different origins of presumed staminodes; pollen grain diameter, in vitro pollen germination and regression between grain diameter and in vitro pollen germination in pickerelweed (pontederia cordata L.) and the development of novel pollination techniques to overcome the effects of heteromorphic incompatibility and herkogamy in pickerelweed.

Chapter 1 - Asteraceae is a large family in the Plant Kingdom. The many flowers of Asteraceae have been cultivated for their desirable esthetics all over the world, and are also used medicinally. In particular, their anti-inflammatory properties are considered to be useful. The active components in numerous flowers of Asteraceae have been identified as flavonoids and their glycosides, sterols, triterpenoids and their glycosides and sesquiterpenes. With regard to flavonoids, various compounds have been isolated from pot marigold, wormwood, edible chrysanthemum, mountain arnica and French marigold, and the chemical compositions have been identified. Most have an anti-inflammatory effect. Triterpenoids and their glycosides have been isolated from artichoke flower, pot marigold, edible chrysanthemum, mountain arnica and sunflower, and the chemical compositions have been identified. These compounds show anti-inflammatory activity and cancer-preventive effects. As components in dietary and herbal supplements, these compounds are considered to be relatively non-toxic to humans. Alkanediols are newly discovered compounds and have been confirmed to show anti-inflammatory and cancer-preventive effects. In this chapter, the authors provide an outline of the bioactive compounds present in the flowers of Asteraceae.

Chapter 2 - The *Arachis* genus is divided into 9 taxonomic sections by Krapovicaks and Gregory (1994) based on the most important morphological characters. It is estimated that there are some 100 species within the genus. In the past, phylogenetic relationships were mainly inferred from morphological features, crossability, karyotype analysis and protein markers; in recent years, however, newer DNA markers and *in situ* hybridization techniques have been used in systematic studies of the genus, resulting in findings somewhat different from previously reported postulations. Globally, four wild peanut species have been used as ornamental groundcover. In Florida, USA, rhizoma perennial peanut, *A. glabrata*, is considered a good sod alternative in the urban landscape. In North Queensland, Australia, *A. pintoi* is used for bank stabilization and as ornamental groundcover in shaded and erosion-prone regions; in South China, the species has been extensively planted in orchards and gardens. *A. duranensis* is accepted as a good ornamental groundcover in South China; it also works well when grown in a hanging pot. *A. repens* is widely utilized as an ornamental both along roadsides and in home yards in South America. Up to now, most of the studies related to wild peanut species have been conducted by peanut breeders, rather than workers on flowers and turf science, with aims to enhance the productivity and stress resistances of *A. hypogaea* more often and in some cases to transfer desirable quality traits from wild peanut species. Research on utilization of wild *Arachis* species in breeding ornamental groundcover cultivars should be further strengthened.

Chapter 3 - Betalains, anthocyanins, carotenoids and chlorophylls, are the main pigments present in flowers. Anthocyanins and betalains are mutually exclusive and they are never found together in a plant. The main function of the pigments in flowers is giving them colour and thus to provide an attraction signal to pollinators. The hue of flowers, however, not only depends on the nature of their pigments but it is also influenced by the presence of co-pigments, the cellular pH and the incorporation of metal ions. Other factors like light conditions and temperature also affects the spectral distribution of the light reflected by the pigments and as consequence the flower color. Several auhors have also observed fluorescence emission from betalains in flowers. The reflected light and the light emitted as fluorescence by flowers pigments may be analysed in terms of their relevance in biosignalling towards pollinators. For this survey, not only the spectral distribution of the light given off from the flowers, but also the sunlight spectrum and the sensitive response of pollinators photoreceptors should be considered. Some flowers pigments show also a defensive function acting as toxics to predator insects. It was found that carotenoids are related to flowers flavor as they act as precursors of chemicals which are responsible for the fragrance of some flowers.

Chapter 4 - Innovation, sustainability and safety of drugs have become the main foci of the modern pharmaceutical industry. Consumer awareness of the possible side effects of the use of chemical-based therapeutic agents has compelled researchers to probe and explore natural botanical-based agents that are toxicologically safe, especially when used in the health care system. To this effect, alternative sources of drugs are being probed for novel lead molecules in an endeavour to offer permanent and sustainable solutions to patients. Plants with potential therapeutic value have been used since time immemorial to cure various ailments and it is only during the last past decades that the scientific community has expressed renewed interest in plants as alternative therapeutic sources of lead molecules. Flowers have traditionally been used in cooking in various cultures, such as European, Asian, East Indian, Victorian English, and Middle Eastern. Being an important part of a plant, flower as any botanical source possess secondary metabolites or the bioactive compounds (phytochemicals) which have been reported to be accountable for various observed biological activities and health benefits.

Phytochemicals produced from flowers exhibit pharmacological effects on the body ranging from anti-inflammatory and antimicrobial properties to cardiovascular benefits. This book chapter focuses on providing and updating available information on the therapeutic activities exhibited by common flowers (Saffron, Lavender, Chamomile, Citrus and Calendula), which are envisaged to find potential applications as alternative natural preservatives for foods and/or applications in the pharmaceutical industries to develop new and sustainable botanical-based products for treating/managing various human ailments.

Chapter 5 - *Sauvagesia erecta* L. is a small pantropical herb in the family Ochnaceae. Flowers are unusual in that in addition to petals there is an inner whorl of five petaloid structures and clusters of small spathulate appendages situated between the petals and petaloid structures. The petaloid structures overlap to form a cone enclosing the androecium and gynoecium. Both the spathulate appendages and petaloid structures have traditionally been considered to be staminodial in origin.

The floral development and floral anatomy of *S. erecta* was investigated using scanning electron microscopy and light microscopy. The author's analysis shows the late initiation of antepetalous petaloid structures following petals, gynoecium, and antesepalous stamens. This is consistent with the phyllotactic and developmental sequence for a second whorl of stamens. Further to this the vascular traces feeding both the petaloid structures and the stamens split from a common trace low in the receptacle of the flower. These observations confirm the staminodial nature of these petaloid structures. The primordia for the spathulate appendages arise very late in the ontogeny of the flower after all other organs have initiated and are already well developed. The receptacle between the petals and petaloid staminodes expands into a narrow androgynophore and the initial spathulate appendages arise in an antesepalous position below the stamens and staminodes. Further primordia develop around the first-formed appendages in centrifugal sequence, eventually joining neighbouring appendages in a continuous girdle. The vascular traces associated with the appendages are randomly attached to the staminal traces and divide repeatedly before connecting to individual structures.

The nature of a corona is discussed and it is concluded that the generally used definition is not adequately reflecting the homology of structures described as a corona. The author's findings strongly suggest that the appendages are novel organs and could be considered as equating to the corona structures seen for example in Passifloraceae, Velloziaceae or Amaryllidaceae. Therefore the appendages are best interpreted as hypanthial emergences and not as staminodes. Comparison with other Ochnaceae suggests that outer filamentous staminodes found in several genera are best interpreted as a corona, in contrast to inner antepetalous staminodes. However, the function of the appendages in plant-insect interactions is unclear, except for attraction. Flowers of *Sauvagesia* are buzz-pollinated and the cone-like petaloid staminodes would appear to have a dual function in protecting the reproductive organs from damage and in directing the pollen released onto visitors.

Chapter 6 - Pickerelweed (*Pontederia cordata* L.) is a tristylous species that utilizes heteromorphic incompatibility to reduce or prevent self-pollination. Three distinct floral morphs are produced by tristylous species, but each plant always produces flowers of the same morph. Previous reports have suggested that pollen produced by the three sets of anthers in pickerelweed differed from one another in grain diameter and in the length of pollen tubes generated during in vivo germination. A correlation has also been described between these variables, which suggests that pollen storage reserves play a role in compatibility of some combinations. The objective of this experiment was to verify previously reported grain diameter data and to determine whether in vitro pollen tube growth is influenced by grain diameter. Analysis of pollen from 12 plants (four each of S-morph, M-morph and L-morph) revealed that diameters of pollen grains produced by anthers borne by the three filament lengths of pickerelweed were significantly different from one another. Diameters of grains of s-pollen averaged 20.46 ± 0.34 μm, while mean diameters of m-pollen and l-pollen measured

35.04 $\pm$ 0.49 μm and 44.97 $\pm$ 0.34 μm, respectively. No overlap in grain diameter occurred among the three classes of pollen. Pollen tubes produced in vitro by l-pollen and m-pollen averaged 486.43 μm and 431.14 μm in length, respectively, 240 min after germination, while pollen tubes from s-pollen attained an average length of 265.57 μm. Previous reports suggested that pollen tube lengths produced in vivo by the three pollen diameter classes were significantly different from one another; however, the author found no difference between lengths of pollen tubes from l-pollen and m-pollen produced during in vitro germination. The reason for these conflicting results is unknown but it is possible that other factors influence in vivo germination. A significant positive regression between pollen grain diameter and in vitro pollen tube length was identified; these results are similar to those described by other workers for in vivo pollen germination and suggest that pollen grain diameter has a positive influence on the length of pollen tubes produced during in vitro germination. This research provides evidence that pollen grain size and tube length may contribute to self-incompatibility in some, but not all, morph interactions in pickerelweed.

Chapter 7 - Pickerelweed is a primarily outcrossing tristylous species that utilizes herkogamy to reduce the likelihood of self-pollination. In addition to this physical separation of reproductive organs, pollen grains borne at the three positions within tristylous flowers produce tubes that correspond in length to reciprocally positioned styles. Most genetic studies require that plants be self-pollinated in order to determine gene action controlling traits and inbreeding can also be useful for developing inbred lines that may have utility in cultivar creation programs. The three morphs of pickerelweed have varying levels of self-incompatibility; M-morphs are self-fertile, whereas S- and L-morphs exhibit reduced fertility after normal self-pollination. In this study the author developed techniques to reduce self-incompatibility in the S- and L-morphs of pickerelweed. Seed set in S-morphs self-pollinated using normal pollen transfer averaged 19.7%, whereas corolla removal increased seed production in these plants to an average of 43.6%. Seed set in L-morphs that were self-pollinated using normal methods averaged 2.7%, whereas stylar surgery of these plants increased average seed production to 29.4%. The methods developed in these experiments will be helpful for plant breeders and geneticists interested in studying this and other tristylous species.

ISBN: 978-1-62808-798-7
© 2013 Nova Science Publishers, Inc.

Chapter 1

Bioactive Components from Asteraceae Flowers

Ken Yasukawa[*]

Nihon University School of Pharmacy,
Narashinodai, Funabashi-shi, Chiba, Japan

Abstract

Asteraceae is a large family in the Plant Kingdom. The many flowers of Asteraceae have been cultivated for their desirable esthetics all over the world, and are also used medicinally. In particular, their anti-inflammatory properties are considered to be useful. The active components in numerous flowers of Asteraceae have been identified as flavonoids and their glycosides, sterols, triterpenoids and their glycosides and sesquiterpenes. With regard to flavonoids, various compounds have been isolated from pot marigold, wormwood, edible chrysanthemum, mountain arnica and French marigold, and the chemical compositions have been identified. Most have an anti-inflammatory effect. Triterpenoids and their glycosides have been isolated from artichoke flower, pot marigold, edible chrysanthemum, mountain arnica and sunflower, and the chemical compositions have been identified. These compounds show anti-inflammatory activity and cancer-preventive effects. As components in dietary and herbal supplements, these compounds are considered to be relatively non-toxic to humans. Alkanediols are newly discovered compounds and have been confirmed to show anti-inflammatory and cancer-preventive effects. In this chapter, we provide an outline of the bioactive compounds present in the flowers of Asteraceae.

Abbreviations

ABTS:	2,2'-Azino-bis-(3-ethylbenzothiazoline-6-sulfonic acid
COX-2:	Cyclooxigenase-2
DMBA:	7,12-Dimethylbenz[*a*]anthracene

---

[*] E-mail: yasukawa.ken@nihon-u.ac.jp, yasukawa.ken@nihon-u.ne.jp.

DPPH:	1,1-Diphenyl-2-picrylhydrazyl
EBV-EA:	EBV-early antigen
FRAP:	Ferric reducing antioxidant power
HDL:	High-density lipoprotein
HMG-CoA:	3-Hydroxy-3-methylglutaryl-coenzyme A
HO-1:	Heme oxygenase-1
HPLC:	High performance liquid chromatography
IC50:	50% Inhibitory concentration
ICAM-1:	Intercellular adhesion molecule-1
ID50:	50% Inhibitory dose
IL-1β, -6:	Interleukin-1β, Interleukin-6
iNOS:	Inducible NO synthase
IR:	Infra red
LDA:	Lithium diisopropylamide
LDL:	Low-density lipoprotein
LPS:	Lipopolysaccharide
MABA:	Microplate alamar blue assay
MIC:	Minimum inhibitory concentration
MIT:	Microculture tetrazolium
MPO:	Myeloperoxidase
MS:	Mass spectrometry
MTPA:	α-Methoxy-α-(trifluoromethyl)phenylacetic acid
Myd88:	Myeloid differentiation factor 88
NF-κB:	Nuclear factor-κB
NMR:	Nuclear magnetic resonance
NO:	Nitric oxide
NOS	NO synthase
NSAIDs:	Nonsteroidal anti-inflammatory drugs
Oct-1:	Octamer transcription factor-1
p38 MARK:	p38 Mitogen-activated protein kinase
PGE2:	Prostaglandin E_2
PTP1B:	Protein tyrosine phosphatase 1B
STAT5:	Signal transducer and activator of transcription 5
TBP:	TATA -binding protein
TLR-4:	Toll-like receptor 4
TNF-α:	Tumor necrosis factor-α
TPA:	12-O-Tetradecanoylphorbol-13-acetate
UV:	Ultra violet

1. INTRODUCTION

Asteraceae (commonly referred to as the aster, daisy or sunflower family) comprise an exceedingly large and widespread family of Angiospermae [1]. The group has more than 23,000 currently accepted species, spread across 1,620 genera and 12 subfamilies. The main feature of the family is the composite flower type in the form of a capitulum surrounded by involucral bracts. Asteraceae have remarkable ecological and economic importance, and are present from the polar regions to the tropics, colonizing all available habitats. Most members of Asteraceae are herbaceous, but a significant number are also shrubs, vines and trees. The family has a worldwide distribution, and is most common in the arid and semi-arid regions of subtropical and lower temperate latitudes. Asteraceae are an economically important family, and are used in numerous herbal medicines, including *Grindelia*, *Echinacea* and yarrow. The most evident characteristic of Asteraceae is their inflorescence; a specialized capitulum, technically called a calathid or calathidium, but generally referred to as flower head. Commercially important plants of Asteraceae include the food crops *Lactuca sativa* (lettuce), *Cichorium* (chicory), *Cynara scolymus* (globe artichoke), *Helianthus annuus* (sunflower), *Smallanthus sonchifolius* (yacon), *Carthamus tinctorius* (safflower) and *Helianthus tuberosus* (Jerusalem artichoke). Many members of the family are grown as ornamental plants for their flowers and some are important ornamental crops for the cut flower industry. Other commercially important species are used as herbs and in herbal teas and other beverages; for example, chamomile, which comes from two different species, the annual *Matricaria recutita* or German chamomile, and the perennial *Chamaemelum nobile*, also called Roman chamomile. *Calendula*, also known as pot marigold, is grown commercially for herbal teas and the potpourri industry. Echinacea (*Echinacea purpurea*) is used as a medicinal tea. Winter tarragon, also called Mexican mint marigold (*Tagetes lucida*), is commonly grown and used as a tarragon substitute in climates where tarragon will not survive. Finally, the wormwood genus *Artemisia* includes absinthe (*A. absinthium*) and tarragon (*A. dracunculus*). Common in all commercial poultry feed, marigold (*Tagetes patula*) is grown primarily in Mexico and Central American nations. Plants in the Asteraceae are also medically important in areas that do not have access to Western medicine. Many members of Asteraceae are copious nectar producers and are useful for evaluating pollinator populations during their bloom; *Centaurea* (knapweed), *Helianthus annuus* (sunflower), and some species of *Solidago* (goldenrod) are major "honey plants" for beekeepers, while the genera *Chrysanthemum*, *Pulicaria*, *Tagetes* and *Tanacetum* contain species with useful insecticidal properties.

2. FLOWERS OF ASTERACEAE

2.1. Capitula of Asteraceae

The capitula are typically solitary or arranged in corymbose or paniculate synflorescences. Corymbose synflorescences are flat-topped, whereas paniculate synflorescences are rounded or pyramidal. Flowering sequences among the capitula vary, and the synflorescences frequently contain combinations of cymose and racemose parts. Only when all capitula are developed in a distinctly cymose sequence are they described as cymose rather than corymbose.

Heterogamous capitula contain florets with different sex arrangements (usually female and hermaphroditic). Homogamous capitula contain florets with similar sex arrangements; they are mostly perfect (i.e., hermaphroditic and fertile). Disciform capitula are usually heterogamous and contain female as well as hermaphroditic florets. Discoid capitula are mostly homogamous with all florets perfect, although there are exceptions. Thus, there are a number of genera and groups of genera with monoecious or dioecious capitula.

2.2. Florets of Asteraceae

Morphologically, there are actinomorphic and zygomorphic florets. The actinomorphic florets always are arranged centrally in the capitulum, or, in discoid and most disciform capitula, are present throughout the capitulum. In disciform capitula the two types of florets are referred to as central and outer florets; the latter may be either actinomorphic or more rarely zygomorphic. Actinomorphic disc and central florets are usually 5-lobed, but several genera are consistently 4-lobed, and some species have 3-lobed central florets.

3. ULTRAVIOLET DEFENSE COMPONENTS OF FLOWERS OF ASTERACEAE

The petals of flowers commonly show differential color markings at their base. Such markings have long been known as nectar guides. Located centrally on the flower, near the nectaries, the guides cue pollinating insects to the presence of adjacent food. Nectar guides are frequently invisible to humans. Consisting of ultra-violet-absorbing patches, they are discernible only to insects such as honey bees, whose visual sensitivity extends into the near-ultraviolet region of the solar spectrum. Thompson et al. demonstrated that ultraviolet absorption in the nectar guide of a composite flower, the familiar black-eyed susan (Rudbeckia hirta), is attributable to a mixture of flavonol glucosides [2]. The petals of the black-eyed susan contain three flavonol glucosides: quercetagetin 7-O-glucoside (1), patulitrin (2), and 6,7-dimethoxy-3',4',5-trihydroxyflavone-3-O-glucoside (3). These compounds, which show intense spectral absorption at 340 to 380 nanometers, are restricted in distribution to the petal bases, which are ultraviolet absorbing as a result. The widespread occurrence of flavonols in flowers suggests that these pigments serve specifically for demarcation of ultraviolet petal patterns visible and relevant to insects. Although visibly yellow like manyflavonoids, they constitute the major group of floral pigments whose chief absorption matches the region detected by the ultraviolet receptors of insects.

4. BIOACTIVE AGENTS FROM FLOWERS OF ASTERACEAE

4.1. Screening of Anti-Inflammatory Effects of Flowers of Asteraceae

The extracts of Compositae plants inhibit 12-O-tetradecanoylphorbol-13-acetate (TPA)-induced ear edema in mice (Table 1) [3]. Seventy-five methanol extracts obtained from 53 species in 11 tribes of Asteraceae plants were assayed and their inhibitory ratios were calculated. In general, medicinal and edible plant extracts are more effective inhibitors than those from other materials, and flower extracts are more effective inhibitors than those from other parts. For example, the methanol extracts of the flowers of Cirsium arvense, Taraxacum *officinale* and *T. platycarpum* (dandelion), *Chrysanthemum morifolium* var. *sinensis* forma *esculentum* (edible chrysanthemum), *C. nipponicum*, *Matricaria matricarioides* and *Helianthus annuus* (sunflower) markedly inhibit the inflammation induced by TPA in mice.

Eleven tublar and nine ligulate flowers from 15 species of Asteraceae plants were investigated for their triterpene alcohol constituents [4]. This led to the isolation and identification of 11 triterpene alcohols, as follows: helianol (**12**), taraxasterol (**27**), ψ-taraxasterol (**21**), α-amyrin (**35**), β-amyrin (**29**), lupeol (**38**), taraxerol (**42**), cycloartenol (**4**), 24-methylenecycloartanol (**5**), tirucalla-7,24-dienol (**14**) and dammaradienol (**15**).

Table 1. Inhibitory effect of the methanol extracts from flowers of Asteraceae plants on TPA-induced inflammation in mice

Tribe	Scientific name (English name)	Part	IR (%)	Source
Subfatnily Cichorioideae				
Cardueae	*Echinops ritro*	t-flower	79	A
	Arctium lappa (burdock)	t-flower	78	A
	Carthamus tinctorius (safflower)	t-flower	73	D
	Centaurea cyanus	t-flower	52	E
	C. nigra	t-flower	57	E
	Cirsium arvense (creeping thistle)	t-flower	80	C
	C. nipponicum	t-flower	40	E
	C. tanakae	t-flower	36	E
	Cynara cardanculus (cardoon)	t-flower	60	A
	Silybum marianum (St. Mary's thistle)	t-flower	66	A
Lactuceae	*Cichorium intybus* (cichory)	l-flower	71	A
	Ixeris debilis	l-flower	53	F
	Sonchus oleraceus	l-flower	54	A
	Hypochaeris radicata	l-flower	39	F
	Lapsana communis (nipplewort)	l-flower	49	G
	Taraxacum officinale (dandelion)	l-flower	95	F
	T. platycarpum (dandelion)	l-flower	87	F
Subfamily Asteroideae				
Inuleae	*Inula helenium* (elecampane)	flower	31	A
Gnaphalieae	*Anaphalis margaritaceae* var. *angustior*	f-head	25	A
Calenduleae	*Calendulla officinalis* (pot marigold)	t-flower	43	C
		l-flower	59	C
Astereae	*Aster tataricus* (purple aster)	t-flower	44	A
		l-flower	41	A
	Solidago altissima	spike	60	F
	Bellis perennis (common daisy)	f-head	38	G
		t-flower	38	G
		l-flower	28	G
Anthemideae	*Chrysanthemum boreale*	f-head	55	A
	C. leucanthemum (French chrysanthemum)	t-flower	31	E
		l-flower	58	E

Table 1. (Continued)

Tribe	Scientific name (English name)	Part	IR (%)	Source
Anthemideae	*Chrysanthemum makinoi*	t-flower	36	A
		l-flower	50	A
	C. morifolium var. *sinensis* (chrysanthemum)	t-flower	39	B
		l-flower	79	B
	C. morifolium var. *sinensis* forma *esculentum* (edible chrysanthemum)	l-flower	87	B
	C. nipponicum	t-flower	82	A
		l-flower	22	A
	C. pacificum	t-flower	44	A
	Matricaria matricarioides	t-flower	58	A
		l-flower	92	
	M. recutita (German chamomile)	f-head	68	A
	Tanacetum vulgare	t-flower	72	G
		f-head	37	G
Senecioneae	*Senecio vulgaris*	l-flower	62	G
	Farfugium japonicum	t-flower	36	A
		l-flower	74	A
Helenieae	*Tagetes patula* (French marigold)	t-flower	45	A
		l-flower	39	A
Heliantheae	*Heliopsis helianthoides*	l-flower	75	A
	Spilanthus acmella var. *oleracea*	t-flower	45	A
	Helianthus annuus (sunflower)	t-flower	92	A
		l-flower	84	A
	H. tuberosus	t-flower	31	F
		l-flower	73	F
	Cosmos bipinnatus (cosmos)	t-flower	67	A
		l-flower	73	A
Eupatorieae	*Eupatorium perfoliatum*	flower	81	C

The samples were applied 30 min before TPA treatment; ear thickness was determined at 6 h after TPA treatment. l-flower, ligulate flower; t-flower, tabular flower; f-head, flower head. A, cultivated at Medicinal Plant Garden in School of Pharmacy, Nihon University; B, purchased at market in Tokyo; C, purchased at herb market in London; D, cultivated at Mogami, Yamagata, Japan; E, collected locally in Nagano, Japan; F, collected locally in Chiba, Japan; G, collected locally in London. Yasukawa, K., *et al.*, *Phytother. Res.*, 1998, 12, 484−487 [3].

The tublar flowers of Calendula officinalis, Carthamus tinctorius, Cosmos bipinnatus, Chrysanthemum morifolium, Helianthus annuus and Matricaria matricarioides contained helianol (**12**) as the most predominant component (29−86%) in the triterpene alcohol fractions. The triterpene alcohols from Asteraceae flowers were evaluated with respect to their anti-inflammatory activity against TPA-induced inflammation (1 μg/ear) in mice. All of these showed marked inhibitory activity, and their 50% inhibitory dose (ID_{50}) was 0.1~0.8 mg/ear.

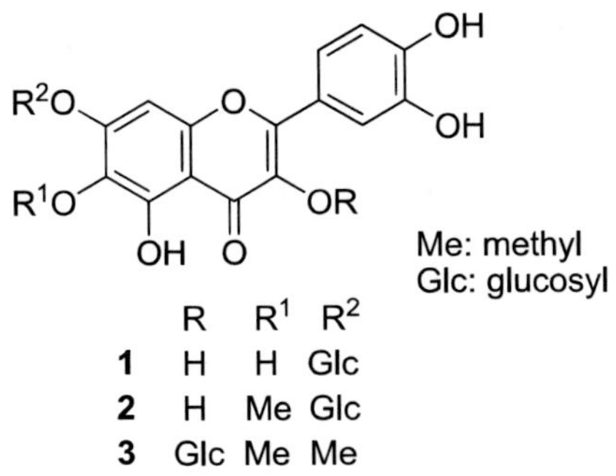

Figure 1. Structures of flavonoids from flowers of *Rudbeckia hirta*.

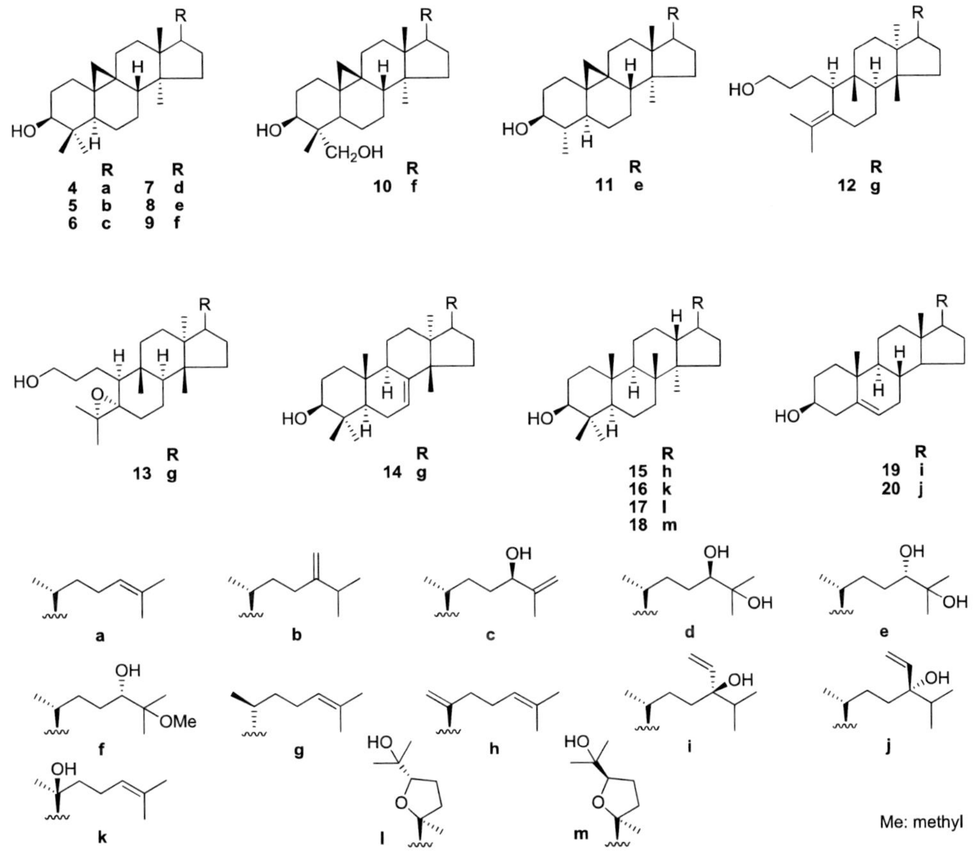

Figure 2. Structures of tetracyclic triterpenoids from edible chrysanthemum.

Ten dihydroxy- and trihydroxy-triterpenes, comprising four taraxastanes (faradiol (**22**), heliantriol B$_0$ (**25**), heliantriol C (**23**) and arnidiol (**28**)), two lupanes (calenduladiol (**40**) and heliantriol B$_2$ (**41**)), two oleananes (maniladiol (**30**) and longispinogenin (**32**)) and two ursans (brein (**36**) and uvasol (**37**)), isolated from the nonsaponifiable lipids of the flower extracts of Asteraceae plants were evaluated with respect to their anti-inflammatory activity against TPA-induced inflammation in mice [5]. All were found to possess marked inhibitory activity. The ID$_{50}$ of these compounds with respect to TPA-inflammation (1 μg) was 0.03~0.2 mg/ear.

Twenty-eight 3-hydroxy triterpenoids, comprising seven taraxastane, five oleanane, three ursane, four lupane, one taraxane, five cycloartane, three tirucallane and one dammarane types isolated from the non-saponifiable lipid fraction of the flower extract of *Chrysanthemum morifolium* were tested for their antitubercular activity against *Mycobacterium tuberculosis* strain H37Rv using the Microplate Alamar Blue Assay (MABA) [6]. Fifteen compounds showed a minimum inhibitory concentration (MIC) in the range of 4~64 μg/ml, among which maniladiol (**30**; MIC 4 μg/ml), 3-epilupeol (**39**; 4 μg/ml) and 4,5α-epoxyhelianol (**13**; 6 μg/ml) exhibited the highest activity. Cytotoxicity of 3-epilupeol (**39**) against Vero cells gave a 50% Inhibitory concentration (IC_{50}) value of over 62.5 μg/ml, suggesting some degree of selectivity for *M. tuberculosis*.

The triterpenoids taraxastane, oleanane, ursane, lupane, taraxane, cycloartane, dammarane and tirucallane isolated from flowers of Asteraceae plants have previously been reported to exhibit anti-inflammatory effects and are variously competitive and non-competitive inhibitors of the serine proteases trypsin and chymotrypsin [7]. The general features of these triterpenoid protease inhibitors are a hydroxyl group and an appropriate side chain in the region of the molecule distal to the 3-hydroxy group. However, fatty acid esterification of the triterpenoid 3-hydroxy group has a marked effect on inhibitor effectiveness. This suggests a means of rapid alteration of the plant defensive complement *in vivo* and of the bioactivity of these anti-inflammatory compounds.

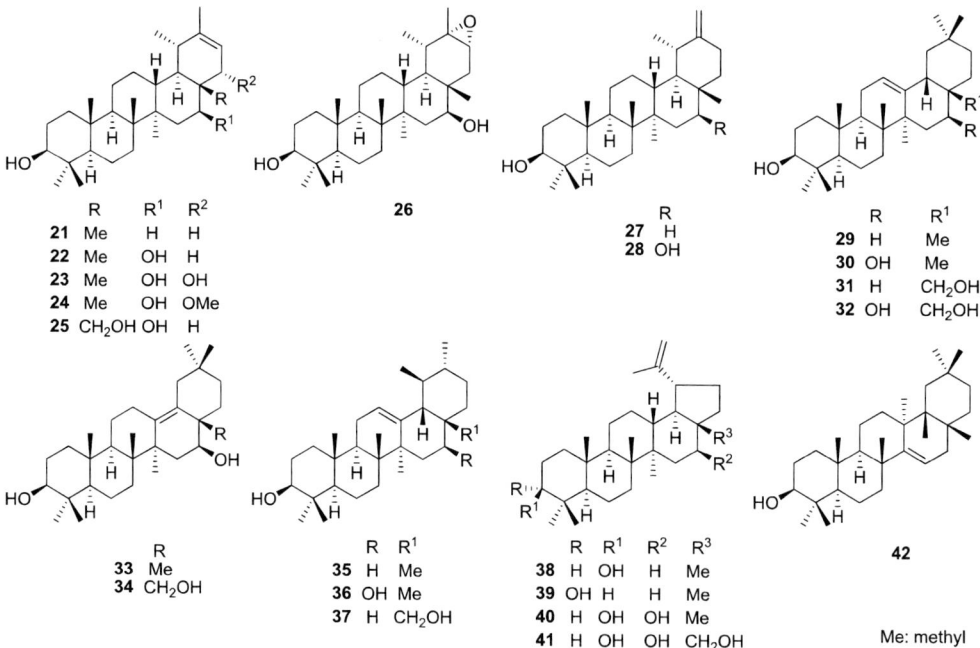

Figure 3. Structures of pentacyclic triterpenoids from edible chrysanthemum.

4.2. Artichoke

The globe artichoke (*Cynara cardunculus* var. *scolymus*) is a variety of thistle cultivated as a food [8]. The edible parts are the buds that form within the flower heads before the flowers come into bloom. The buds disappear or become a coarse, barely edible form when the flower blooms. The uncultivated or wild variety of the species is called a cardoon. It is a perennial plant native to the Mediterranean region. The total antioxidant capacity of artichoke flower heads is one of the highest reported for vegetables [9]. Cynarine (**43**) is a chemical constituent in *Cynara*, and the majority of the cynarine found in artichoke is located in the pulp of the leaves, although it also present in the dried leaves and stems. It inhibits taste receptors, making water (and other foods and drinks) seem sweet. Studies have also shown that artichoke aids digestion, hepatic [10, 11] and gall bladder function [12], and raises the ratio of high-density lipoprotein (HDL) to low-density lipoprotein (LDL) [13]. This reduces cholesterol levels, which diminishes the risk for arteriosclerosis and coronary heart disease [14]. Furthermore, aqueous extracts from artichoke leaves have been shown to reduce cholesterol by inhibiting 3-hydroxy-3-methylglutaryl-coenzyme A (HMG-CoA) reductase and having a hypolipidemic influence, lowering blood cholesterol [15]. Artichoke contains the bioactive agents apigenin (**44**) and luteolin (**45**), and *C. scolymus* seems to have a bifidogenic effect on beneficial gut bacteria [16]. Artichoke leaf extract has proven to be helpful for patients with functional dyspepsia [17], and may ameliorate the symptoms of irritable bowel syndrome [18, 19].

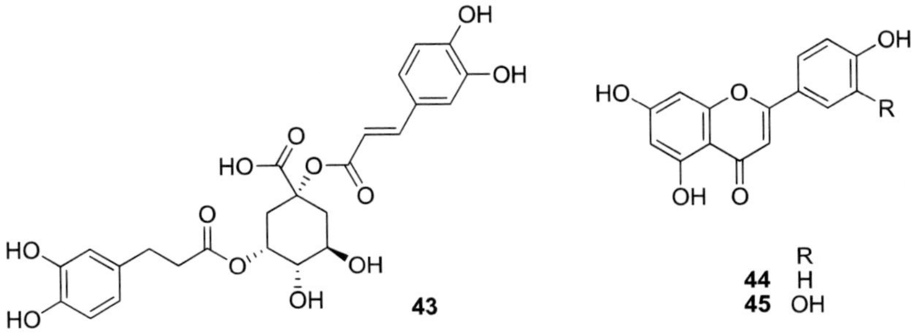

43

	R
44	H
45	OH

Figure 4. Structures of cynarine (**43**) and flavones from artichoke flower.

The methanol extract of the flowers of artichoke exhibited marked antitumor activity in an in vivo two-stage carcinogenesis test in mice, using 7,12-dimethylbenz[*a*]anthracene (DMBA) as an initiator and TPA as a promoter [20]. From the active fraction of the methanol extract, four triterpene alcohols (**21**, **27**, **29**, **35**) and their corresponding acetates (**46~49**) were isolated and identified. These compounds were evaluated for their inhibitory effects on TPA-induced inflammation (1 μg/ear) in mice and showed marked anti-inflammatory effects, with an ID_{50} of 0.50~0.91 μmol/ear.

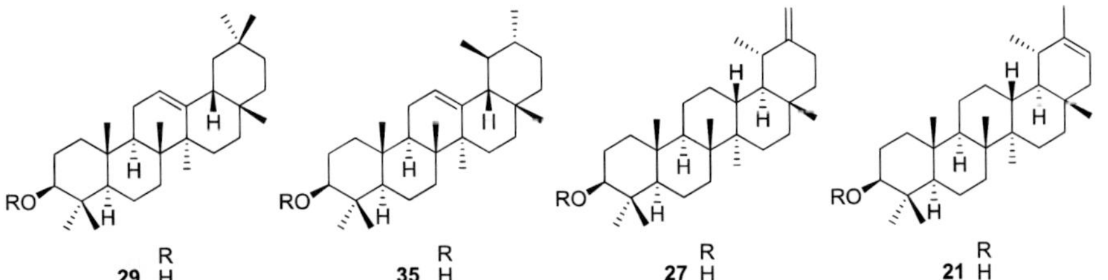

	R
29	H
46	Ac

	R
35	H
47	Ac

	R
27	H
48	Ac

	R
21	H
49	Ac

Ac: acetyl

Figure 5. Structures of triterpenoids from artichoke flower.

4.3. Safflower

Safflower (*Carthamus tinctorius*) is a highly branched, herbaceous, thistle-like annual. It is native to arid environments having seasonal rain, and is one of humanity's oldest crops, being commercially cultivated for vegetable oil extracted from the seeds. Chemical analysis of ancient Egyptian textiles dated to the Twelfth Dynasty identified dyes made from safflower, and garlands made from safflowers were found in the tomb of the pharaoh Tutankhamun [21]. Safflower was also known as carthamine in the nineteenth century. Traditionally, the crop was grown for its seeds, and used for coloring and flavoring foods, in medicines, and making red (carthamin) and yellow dyes, particularly before cheaper aniline dyes became available [21]. High-linoleic safflower oil has been shown to increase adiponectin, a protein that helps regulate blood glucose levels and fatty-acid breakdown. Safflower flowers are occasionally used in cooking as a cheaper substitute for saffron, and were thus often referred to as "bastard saffron". Dried safflower flowers are used as a natural textile dye, and in traditional Chinese medicine to alleviate pain, increase circulation and reduce bruising. They are included in herbal remedies for menstrual pain and minor physical trauma. In India, the flowers are used for their laxative and diaphoretic properties, and are also used for childhood measles, fevers and eruptive skin conditions.

Tumor inhibition of two-stage skin carcinogenesis has been observed with the methanol extract of the safflower flower, which is a traditional Chinese medicine and natural pigment of rouge additives in several Asian countries [22]. From these active fractions, $\Delta 5$- and $\Delta 7$-sterol fractions were separated, and were examined for inhibitory activity against TPA-induced inflammatory ear edema in mice. Stigmasterol (**50**) was the most abundant of 14 sterols identified in the $\Delta 5$-sterol fraction, while schottenol (**51**) constituted the dominant sterol in the $\Delta 7$-sterol fraction. Furthermore, stigmasterol markedly inhibited tumor promotion in two-stage skin carcinogenesis experiments.

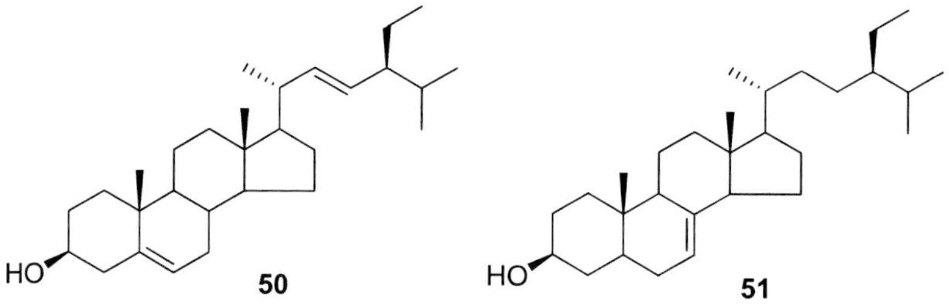

Figure 6. Structures of stigmasterol (**50**) and schottenol (**51**) from safflower.

A C_{31} alkanediol isolated from the methanol extract of safflower flower petals was established to have the structure (6*R*,8*S* and/or 6*S*,8*R*)-*syn*-hentriacontane-6,8-diol (**52**) using spectral and chemical methods [23]. Eleven other *syn*-alkane-6,8-diols with carbon numbers C_{21}- (**53**), C_{23}- (**54**), C_{25}- (**55**), C_{27}-~C_{30}- (**56~59**), C_{32}-~C_{35}- (**60~63**) were also isolated and characterized.

Then, 11 secondary alkane-1,3-diols were isolated from a methanol extract of safflower flower petals [24]. Their structures were determined to be *syn*-(*R,S* and/or *S,R*)-C_{36}-alkane-6,8-diol (**64**), *syn*-C_{28}- (**65**), C_{30}- (**66**), C_{32}- (**67**), C_{34}-(**68**), and C_{36}-alkane-7,9-diols (**69**), and *syn*-C$_{27}$- (**70**), C$_{29}$- (**71**), C$_{31}$- (**72**), C$_{33}$- (**73**), and C$_{35}$-alkane-8,10-diols (**74**) by spectral methods.

The $\Delta\delta$ ($\delta S-\delta R$) values for the C-1 methyl ^{1}H-Nuclear magnetic resonance (^{1}H-NMR) spectroscopy of the bis-α-Methoxy-α-(trifluoromethyl)phenylacetic acid (MTPA) esters of four synthetic stereoisomers of alkane-6,8-diols, viz, bis-MTPA esters of (6*S*,8*R*)-C$_{27}$- and C$_{29}$- ($\Delta\delta$= −0.05 ppm), (6*R*,8*S*)-C$_{27}$- and C$_{29}$- ($\Delta\delta$= +0.05 ppm), (6*S*,8*S*)-C$_{27}$- ($\Delta\delta$= −0.01 ppm), and (6*R*,8*S*)-C$_{27}$- ($\Delta\delta$= +0.01 ppm) alkane-6,8-diols, made it possible to differentiate unequivocally among the four stereoisomers [25]. This allowed determination of the (6*S*,8*R*)-stereochemistry ($\Delta\delta$= −0.05 ppm for the bis-MTPA esters) for the natural C$_{27}$- and C$_{29}$-alkane-6,8-diols isolated from the flowers of three Asteraceae plants, *Carthamus tinctorius*, *Cynara cardunculus*, and *Taraxacum platycarpum*.

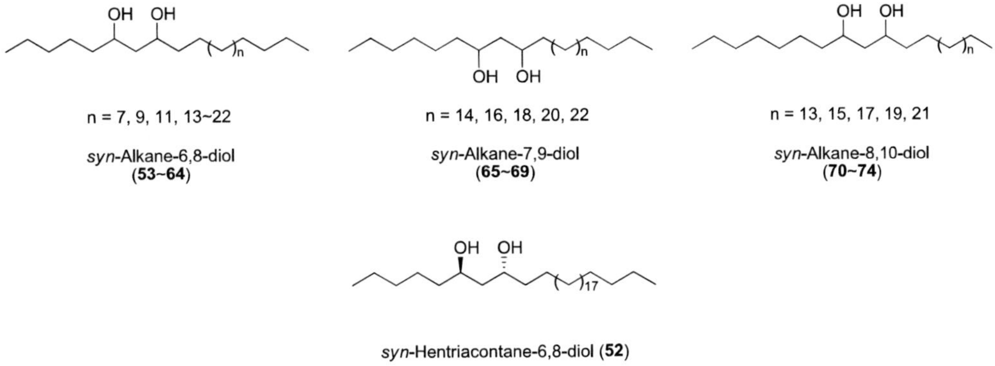

n = 7, 9, 11, 13~22

syn-Alkane-6,8-diol
(**53~64**)

n = 14, 16, 18, 20, 22

syn-Alkane-7,9-diol
(**65~69**)

n = 13, 15, 17, 19, 21

syn-Alkane-8,10-diol
(**70~74**)

syn-Hentriacontane-6,8-diol (**52**)

Figure 7. Structures of aklanediols (52~74) from safflower.

Ten tubular and eight ligulate flowers and seven flower heads from 22 species of Asteraceae were investigated for their alkanediol constituents [26]. All of the flowers contained alkanediols in small amounts suggesting their widespread occurrence in these flowers. Twelve alkanediols were identified as *syn*-(*R*,*S* and/or *S*,*R*)-C$_{21}$- (**53**), C$_{23}$- (**54**), C$_{25}$- (**55**), and C$_{27}$~C$_{35}$-alkane-6,8-diols (**56~63**), among which *syn*-hentriacontane-6,8-diol (**52**) were abundant in many of the flowers.

Homologs C$_{15}$- and C$_{29}$- in the alkane-6,8-diols series were prepared as shown in Figure 8 [27]. The carbanion derived from methyl *p*-tolyl sulfoxide (**A**) with slight excess of lithium diisopropylamide (LDA) was reacted with 1-hexanal to afford a diastereomeric mixture of the 1,2-adduct **B** in 95% yield. In order to obtain 7-(*p*-tolylsulfinyl)-alkane-6,8-diol (**C**), the carbanion derived from 1-(*p*-tolylsulfinyl)heptan-2-ol with **B** equivalents of n-BuLi was reacted with aldehydes to afford the 1,2-adducts in moderate yields. The aldehydes were derived from oxidation of the corresponding alcohols (**Ca~c**) with 2,2,6,6-tetramethylpiperidin-1-yloxy and sodium bromite in good yield. The sulfinyl group of C$_{15}$-, C$_{27}$- and C$_{29}$- was reduced with Raney Ni in ethanol at room temperature to give alkane-6,8-diols (**Da~c**) in moderate yields.

These alkane-6,8-diols were synthesized in moderate yields from a chiral β-ketosulfoxide in three steps, and showed inhibitory activity against TPA-induced ear inflammation in mice [28]. Natural *syn*-alkane-6,8-diols showed a weaker effect when compared with non-natural, synthetic anti-alkane-6,8-diols. Furthermore, the anti-inflammatory effects of these alkane-6,8-diols was highly dependent on the length of the main chain.

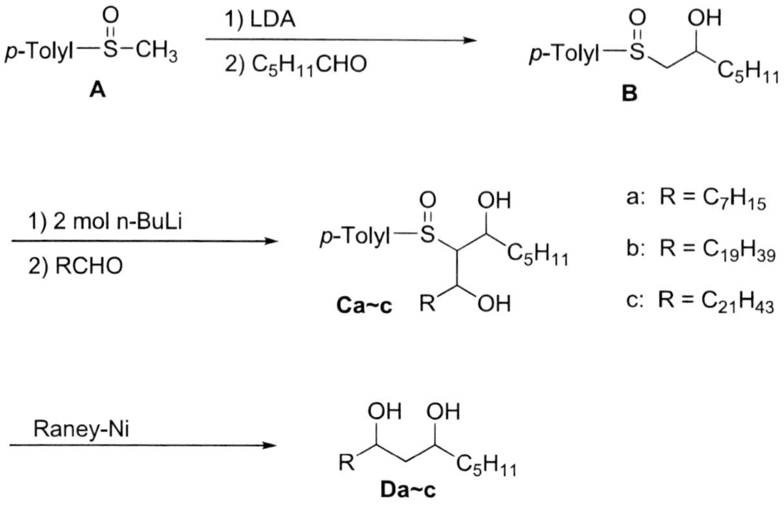

Figure 8. Synthesis of alkane-6,8-diols. (Motohashi, *et al.*, *J. Med. Chem.*, 1995, 38, 4155–4156. [27])

Of eight alkane-6,8-diols assayed, the C_{27}-, C_{31}-, C_{32}-, C_{33}- and C_{35}- alkane-6,8-diols inhibited inflammation in mice. The ID_{50} of these compounds for TPA-induced inflammation was 0.5~0.7 mg/ear. However, C_{21}-, C_{23}- and C_{25}- alkane-6,8-diols were found to have no effect. Furthermore, the mixture of syn-alkane-6,8-diols from safflower flowers markedly suppressed the promotion effect of TPA on skin tumor formation in mice following initiation with DMBA [29].

Hydroxysafflor yellow A (**75**) is an active ingredient obtained from the flower of safflower [30]. The present study investigated the effects of hydroxysafflor yellow A (**75**) on lipopolysaccharide (LPS)-induced inflammatory signal transduction in human alveolar epithelial A549 cells. A549 cells stimulated with LPS were incubated with three doses of hydroxysafflor yellow A (**75**) (1, 4 and 16 μmol/L). Hydroxysafflor yellow A (**75**) suppressed the expression of toll-like receptor 4 (TLR-4), myeloid differentiation factor 88 (Myd88), intercellular adhesion molecule-1 (ICAM-1), Tumor necrosis factor-α (TNFα), Interleukin-1β (IL-1β) and IL-6 at the mRNA and protein levels, and inhibited the adhesion of leukocytes to A549 cells. Hydroxysafflor yellow A (**75**) treatment also decreased nuclear factor-κB (NF-κB) p65 nuclear translocation and inhibited the phosphorylation of p38 mitogen-activated protein kinase (p38 MAPK). These findings suggest that hydroxysafflor yellow A (**75**) effectively inhibits LPS-induced inflammatory signal transduction in A549 cells.

4.3. Pot marigold (*Calendula Officinalis*)

Calendula officinalis (pot marigold, common marigold) is a plant in the genus *Calendula*. It is probably native to southern Europe, though its long history of cultivation makes its precise origin unknown, and it may be of garden origin. It is also widely naturalized further north in Europe (northern and southern England), and elsewhere in warm temperate regions of the world. Pot marigold florets are considered edible. They are often used to add color to salads, or added to dishes as a garnish and in lieu of saffron, and although the leaves are edible, they are often unpalatable.

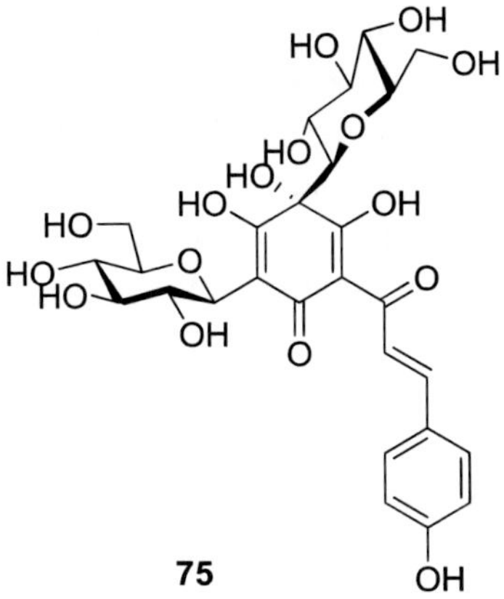

75

Figure 9. Structure of hydroxysafflor yellow A (**75**).

The flowers were used in ancient Greek, Roman, Middle Eastern and Indian cultures as a medicinal herb as well as a dye for fabrics, foods and cosmetics. They are also used to make oil that protects the skin, and many of these uses persist today.

Ten oleanane-type triterpene glycosides, **76~85**, including four new compounds, calendulaglycoside A 6'-*O*-methyl ester (**84**), calendulaglycoside A 6'-*O*-*n*-butyl ester (**85**), calendulaglycoside B 6'-*O*-*n*-butyl ester (**78**), and calendulaglycoside C 6'-*O*-*n*-butyl ester (**82**), along with five known flavonol glycosides, **86~90**, were isolated from the flowers of marigold. Upon evaluation of compounds **76~84** for inhibitory activity against TPA-induced inflammation (1 µg/ear) in mice, all of the compounds, except for **83**, exhibited marked anti-inflammatory activity, with 50% inhibitory dose (ID$_{50}$) values of 0.05~0.20 mg/ear. In addition, when **76~90** were evaluated against the Epstein-Barr virus early antigen (EBV-EA) activation induced by TPA, compounds **76~85** exhibited moderate inhibitory effects (IC$_{50}$ values of 471~487 mol ratio/32 pmol TPA). Furthermore, upon evaluation of the cytotoxic activity against human cancer cell lines in vitro in the NCI Developmental Therapeutics Program, two triterpene glycosides, **76** and **79**, exhibited their most potent cytotoxic effects against colon cancer, leukemia, and melanoma cells.

4.4. Artemisia

Artemisia is a large, diverse genus of plants comprising between 200 and 400 species. Common names for various species in the genus include mugwort, wormwood and sagebrush. *Artemisia* comprises hardy herbaceous plants and shrubs, which are known for the powerful chemical constituents in their essential oils. *Artemisia* species grow in temperate climates of both hemispheres, typically in dry or semiarid habitats. Notable species include *A. vulgaris* (common mugwort), *A. tridentata* (big sagebrush), *A. annua* (sagewort), *A. absinthium* (wormwood), *A. dracunculus* (tarragon), and *A. abrotanum* (southernwood). The leaves of many species are covered with white hairs. Most species have strong aromas and bitter flavors due to terpenoids and sesquiterpene lactones, which exist as an adaptation to discourage herbivory. The small flowers are wind-pollinated. *Artemisia arborescens* (tree wormwood, or sheeba in Arabic) is a very bitter herb indigenous to the Middle East and is used in tea, usually with mint. Artemisinin (**91**), from *Artemisia annua*, is the active ingredient in the antimalarial combination therapy Coartem produced by Novartis and the World Health Organization. Artemisinin and its derivatives are a group of drugs that possess the most rapid action of all current drugs against malaria [32]. Treatments containing a derivative (artemisinin-combination therapies) are now the standard treatment worldwide for *P. falciparum* malaria. The starting compound artemisinin is isolated from the plant *Artemisia annua*. *Artemisia absinthium* (absinthium, absinthe wormwood, wormwood, common wormwood, green ginger or grand wormwood) is native to temperate regions of Eurasia and northern Africa. It is grown as an ornamental plant and is used an ingredient in the spirit absinthe, as well as in bitters, vermouth and pelinkovac. In the middle Ages, it was used to spice mead, while in 18th century England, wormwood was sometimes used instead of hops in beer.

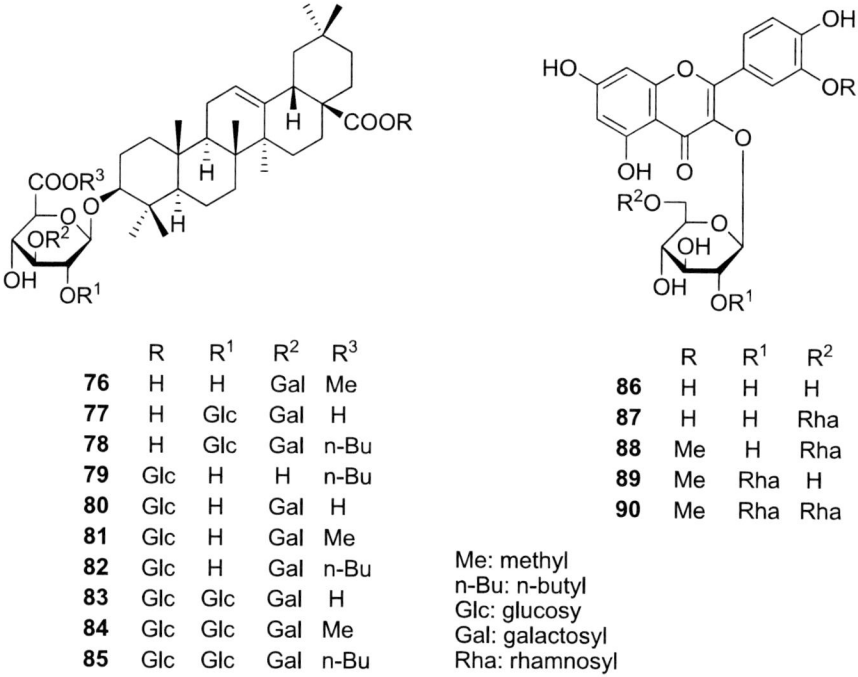

	R	R¹	R²	R³
76	H	H	Gal	Me
77	H	Glc	Gal	H
78	H	Glc	Gal	n-Bu
79	Glc	H	H	n-Bu
80	Glc	H	Gal	H
81	Glc	H	Gal	Me
82	Glc	H	Gal	n-Bu
83	Glc	Glc	Gal	H
84	Glc	Glc	Gal	Me
85	Glc	Glc	Gal	n-Bu

	R	R¹	R²
86	H	H	H
87	H	H	Rha
88	Me	H	Rha
89	Me	Rha	H
90	Me	Rha	Rha

Me: methyl
n-Bu: n-butyl
Glc: glucosy
Gal: galactosyl
Rha: rhamnosyl

Figure 10. Structures of saponins and flavonol glycosides from pot marigold.

91

Figure 11. Structure of artemisinin (**91**) from *Artemisia annua*.

As a part of our ongoing effort to identify anti-diabetic constituents from natural sources, we examined the inhibitory activity of the methanol extracts of 12 species of the genus *Artemisia* against α-glucosidase and protein tyrosine phosphatase 1B (PTP1B) [33]. The methanol extracts of different species exhibited promising α-glucosidase- and PTP1B-inhibitory activities. As the methanol extract of *Artemisia capillaris* exhibited the highest α-glucosidase inhibitory activity together with significant PTP1B inhibitory activity, it was selected for further investigation. Repeated column chromatography based on bioactivity guided fractionation yielded 10 coumarins (umbelliferone (**92**), daphnetin (**93**), 7-methoxy coumarin (**94**), esculetin (**95**), scopoletin (**96**), scoparone (**97**), scopolin (**98**), 6-methoxy artemicapin C (**99**), esculin (**100**), isoesculin (**101**)), 8 flavonoids (linarin (**108**), cirsilineol (**109**), arcapillin (**110**), quercetin (**111**), isorhamnetin (**112**), isorhamnetin 3-glucoiside (**113**), hyperoside (**114**), isorhamnetin 3-robinobioside (**115**)), 6 phenolic compounds (3-caffeoylquinic acid (**102**), 4,5-dicaffeoylquinic acid (**103**), 3,5-dicaffeoylquinic acid (**104**), 3,4-dicaffeoylquinic acid (**105**), 1,5-dicaffeoylquinic acid (**106**), 3,5-dicaffeoylquinic acid methyl ester (**107**)), and one chromone (capillarisin (**116**)). Among these compounds, esculetin (**95**), scopoletin (**96**), quercetin (**111**), hyperoside (**114**), isorhamnetin (**112**), 3,5-dicaffeoylquinic acid methyl ester (**107**), 3,4-dicaffeoylquinic acid (**105**), and 1,5-dicaffeoylquinic acid (**106**) exhibited potent α-glucosidase inhibitory activity when compared to the positive control acarbose. In addition, esculetin (**95**) and 6-methoxy artemicapin C (**116**) displayed PTP1B inhibitory activity.

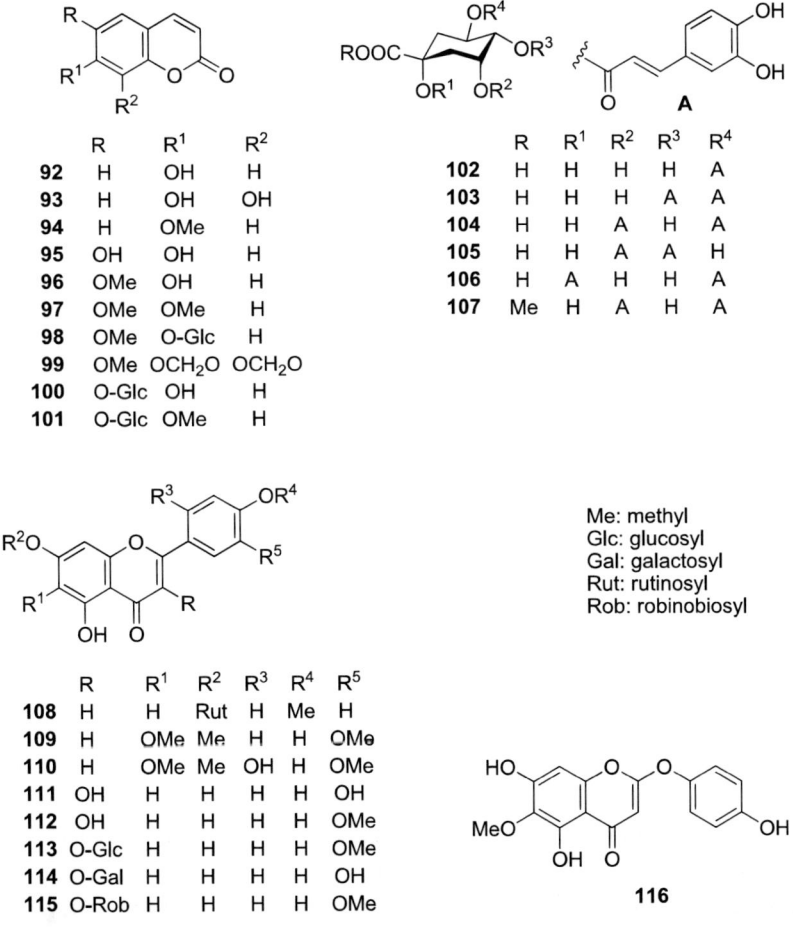

	R	R¹	R²
92	H	OH	H
93	H	OH	OH
94	H	OMe	H
95	OH	OH	H
96	OMe	OH	H
97	OMe	OMe	H
98	OMe	O-Glc	H
99	OMe	OCH₂O	OCH₂O
100	O-Glc	OH	H
101	O-Glc	OMe	H

	R	R¹	R²	R³	R⁴
102	H	H	H	H	A
103	H	H	H	A	A
104	H	H	A	H	A
105	H	H	A	A	H
106	H	A	H	H	A
107	Me	H	A	H	A

	R	R¹	R²	R³	R⁴	R⁵
108	H	H	Rut	H	Me	H
109	H	OMe	Me	H	H	OMe
110	H	OMe	Me	OH	H	OMe
111	OH	H	H	H	H	OH
112	OH	H	H	H	H	OMe
113	O-Glc	H	H	H	H	OMe
114	O-Gal	H	H	H	H	OH
115	O-Rob	H	H	H	H	OMe

Me: methyl
Glc: glucosyl
Gal: galactosyl
Rut: rutinosyl
Rob: robinobiosyl

116

Figure 12. Structures of components from Artemisia spp.

Interestingly, all isolated dicaffeoylquinic acids showed significant PTP1B inhibitory activity. The results of this clearly demonstrated the potential of the *A. capillaris* extract to inhibit α-glucosidase and PTP1B. These inhibitory properties can be largely attributed to a combination of different chemical structures, including coumarins, flavonoids and dicaffeoylquinic acids, which could be further explored to develop therapeutic or preventive agents for the treatment of diabetes.

4.5. Edible Chrysanthemum

Chrysanthemums are perennial flowering plants of the genus *Chrysanthemum* and are native to Asia and northeastern Europe. About 30 species have been described. Yellow or white chrysanthemum flowers of the species *C. morifolium* are boiled to make a sweet drink in some parts of Asia. The resulting beverage is known simply as "chrysanthemum tea" in Chinese. In Korea, a rice wine flavored with chrysanthemum flowers is called gukhwaju. Chrysanthemum leaves are steamed or boiled and used as greens, particularly in Chinese cuisine. Other uses include adding the petals of chrysanthemum to thick snake meat soup in order to enhance aroma. Small chrysanthemums are used in Japan as a sashimi garnish. Chrysanthemums were first cultivated in China as a flowering herb as far back as the 15th century BC. The flower may have been brought to Japan in the eighth century AD. The flower was brought to Europe in the 17th century.

The *n*-hexane soluble and nonsaponifiable lipid fractions of the edible flower extract of chrysanthemum (*Chrysanthemum morifolium*) were investigated for triterpene diol and triol constituents [34]. These triterpenes occur as 3-*O*-fatty acid esters in the *n*-hexane soluble fraction, from which 26 new and 6 known fatty acid esters were isolated and characterized. From the nonsaponifiable lipid fraction, 24 triterpene diols and triols were isolated, of which 3 were new compounds: (24*S*)-25-methoxycycloartane-3β,24-diol (**9**), (24*S*)-25-methoxycycloartane-3β,24,28-triol (**10**), and 22α-methoxyfaradiol (**24**). Faradiol (**22**) and heliantriol C (**23**), present in the nonsaponifiable lipid fraction and as the 3-*O*-palmitoyl esters (**117~125**) in the *n*-hexane soluble fraction, were the most predominant triterpene diol and triol constituents. Fourteen triterpene diols and triols and 9 fatty acid esters were evaluated with respect to their anti-inflammatory activity against TPA-induced inflammation in mice. All of the triterpenes examined showed marked inhibitory activity, with an ID_{50} of 0.03~1.0 mg/ear, which was more inhibitive than quercetin (ID_{50} = 1.6 mg/ear), a known inhibitor of TPA-induced inflammation in mice.

Fifteen pentacyclic triterpene diols and triols, comprising six taraxastanes (faradiol (**22**), heliantriol C (**23**), 22α-methoxyfaradiol (**24**), heliantriol B_0 (**25**), faradiol α-epoxide (**26**), and arnidiol (**28**)) five oleananes (maniladiol (**30**), erythrodiol (**31**), longispinogenin (**32**), coflodiol (**33**), and heliantriol A_1 (**34**)), two ursanes (brein (**36**) and uvaol (**37**)) and two lupanes (calenduladiol (**40**) and heliantriol B_2 (**41**)), isolated from the non-saponifiable lipid fraction of the edible flower extract of chrysanthemum (*Chrysanthemum morifolium*) were evaluated for their inhibitory effects on EBV-EA activation induced by the tumor promoter [35], TPA, in Raji cells as a primary screening test for anti-tumor promoters. All of the compounds tested showed inhibitory effects against EBV-EA activation with potencies either comparable with or stronger than that of glycyrrhetic acid, a known natural anti-tumor-promoter.

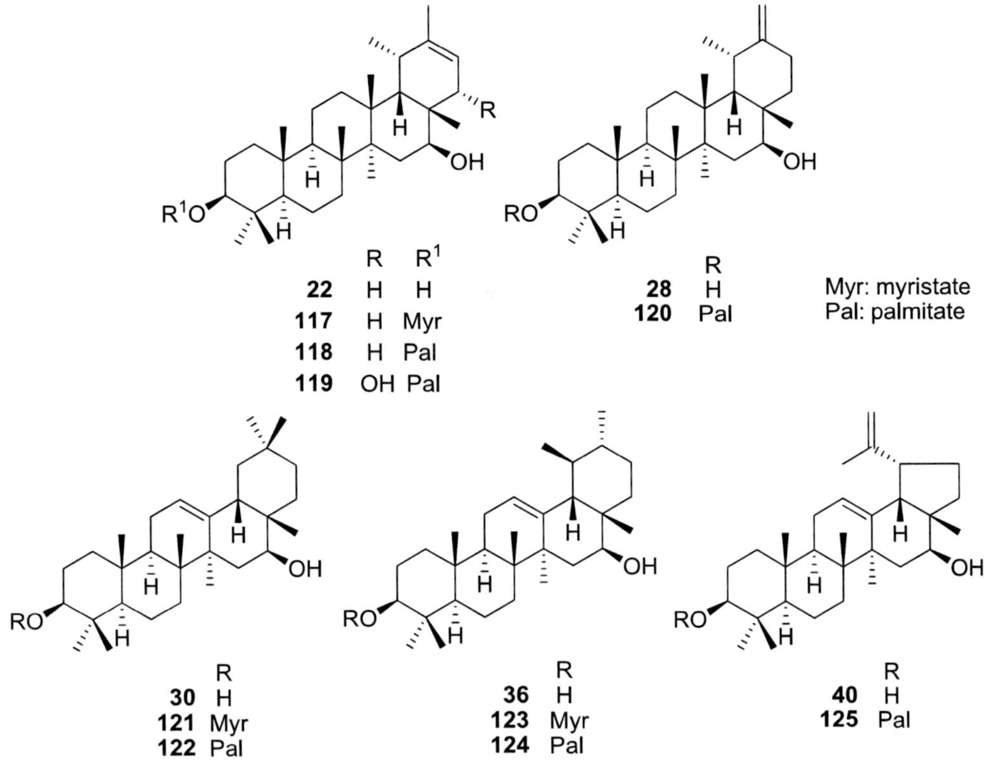

Figure 13. Structures of triterpenoids from edible chrysanthemum.

Evaluation of the cytotoxic activity of six compounds, **22**, **23**, **25**, **26**, **28**, **30**, against human cancer cell lines revealed that arnidiol (**28**) possesses a wide range of cytotoxicity, with GI_{50} values (concentration that yields 50% growth) of mostly less than 6 mM.

Two taraxastane-type hydroxyl triterpenes, taraxasterol (**27**) and faradiol (**22**), isolated from the flowers of Asteraceae plants *Cynara scolymus* (artichoke) and *Chrysanthemum×morifolium* (chrysanthemum), respectively, showed strong inhibitory activity against TPA-induced inflammation in mice [36]. At 2.0 μM/mouse, these compounds markedly inhibited the tumor-promoting effects of TPA on skin tumor formation following DMBA.

Two hydroxyl taraxastane-type triterpenes, faradiol (**22**) and heliantriol C (**23**), were isolated from the ligulate flowers of *Chrysanthemum×morifolium* var. *sinense* farma *esculentum*, the edible chrysanthemum [37]. These compounds showed strong inhibitory activity against TPA-induced inflammation in mice. At 0.2 μM/mouse, these compounds markedly inhibited the promoting effects of TPA on skin tumor formation following DMBA.

Major flavonoids present in the petals of edible chrysanthemum flowers (*Chrysanthemum×morifolium* forma *esculentum*) were identified, and their chemical compounds in relation to their radical scavenging activities and preventive effects against liver injury were compared [38]. Based on retention times and ultra violet (UV) spectra, three peaks on the high performance liquid chromatography (HPLC) chromatogram of the phenolic fraction of edible chrysanthemum flowers confirmed the presence of luteolin (**45**), apigenin 7-O-glucoside (**127**), and luteolin 7-O-glucoside (**128**). Spectroscopic analysis determined the chemical structure of the three newly isolated compounds to be apigenin 7-O-(6"-Omalonyl)-glucoside (**132**), acacetin 7-*O*-(6"-*O*-malonyl)-glucoside (**133**), and luteolin 7-*O*-(6"-*O*-malonyl)-glucoside (**134**). Increases in plasma aspartate aminotransferase and alanine aminotransferase activities in mice (with liver injury induced by injection of carbon tetrachloride) were strongly suppressed by oral administration of luteolin and luteolin 7-*O*-(6"-*O*-malonyl)-glucoside (**134**), which have stronger radical scavenging activity than other compounds. Thus, compounds with chemical structures such as luteolin (**45**) and luteolin 7-*O*-glucoside (**128**), which have malonic acid on its glucosyl moiety, appear to be readily available for mitigation of liver injury.

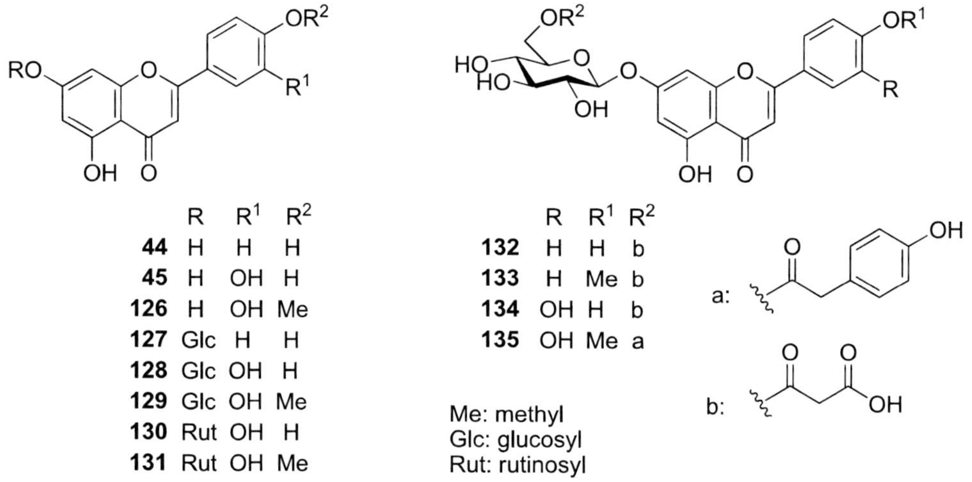

	R	R¹	R²
44	H	H	H
45	H	OH	H
126	H	OH	Me
127	Glc	H	H
128	Glc	OH	H
129	Glc	OH	Me
130	Rut	OH	H
131	Rut	OH	Me

	R	R¹	R²
132	H	H	b
133	H	Me	b
134	OH	H	b
135	OH	Me	a

Me: methyl
Glc: glucosyl
Rut: rutinosyl

Figure 14. Structures of flavonoids from edible chrysanthemum.

A new *p*-hydroxyphenylacetyl flavonoid, diosmetin 7-(6'-*O*-*p*-hydroxyphenylacetyl)-*O*-β-D-glucopyranoside (**135**), was isolated from the flowers of *C. morifolium*, together with five known flavonoids (luteolin (**45**), diosmetin (**126**), diosmetin 7-*O*-D-glucopyranoside (**129**), diosmin (**131**) and scolimoside (**130**)) and four known caffeoylquinic acid derivatives (macranthoin F (**104**), 3,5-dicaffeoylquinic acid (**107**), 1,3-dicaffeoyl-*epi*-quinic acid and chlorogenic acid (**102**)) [39]. The structure of diosmetin 7-(6'-*O*-*p*-hydroxyphenylacetyl)-*O*-β-D-glucopyranoside (**135**) was elucidated by UV, Infra red (IR), NMR spectroscopic methods. Cytotoxic activity of compounds **45**, **126**, **129**, **131**, **135** against human colon cancer cell Colon205 was investigated using MTT assays. Compounds **45** and **126** showed significant cytotoxicity against Colon205, with their IC_{50} values being 96.9 and 82.9 μM, respectively. However, compounds **129**, **131** and **135** showed little cytotoxic activity.

4.6. *Arnica montana* and *A. chamissonis*

Arnica is a genus with about 30 perennial, herbaceous species. This circumboreal and montane (subalpine) genus occurs mostly in the temperate regions of western North America, while two are native to Eurasia (*A. angustifolia* and *A. montana*). Several species, such as *A. montana* and *A. chamissonis*, contain helenalin (**136**), a sesquiterpene lactone that is a major ingredient in anti-inflammatory preparations (used mostly for bruises). *A. montana* has been used medicinally for centuries. *Arnica* is used in liniment and ointment preparations for strains, sprains and bruises. Homeopathic preparations of *Arnica* are also widely marketed and used. In the UK, the Medicines and Healthcare Products Regulatory Agency has registered the product for sprains and bruising under the National Rules for Homoeopathic Products in 2006. *A. montana* is endemic to Europe, from southern Iberia to southern Scandinavia and the Carpathians. It is absent from the British Isles and the Italian and Balkan Peninsulas. *A. montana* grows in nutrient-poor siliceous meadows at elevation up to 3,000 meters (9,800 ft). Helenalin can be poisonous if large amounts of the plant are eaten, producing gastroenteritis and internal bleeding of the digestive tract. When used topically in a gel at 50% concentration, *A. montana* was found to have the same effect when compared to a 5% ibuprofen gel for treating the symptoms of hand osteoarthritis [40].

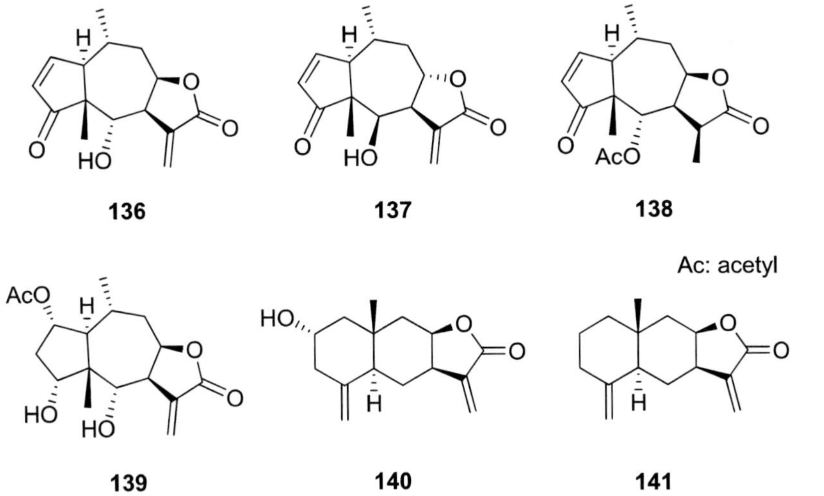

Figure 15. Structure of helenalin (**136**) and its derivatives from arnica flower.

Alcoholic extracts prepared from Arnicae flos, the collective name for flower heads from *Arnica montana* and *A. chamissonis* ssp. *foliosa*, are used therapeutically as anti-inflammatory remedies [41]. The active ingredients mediating the pharmacological effects are mainly sesquiterpene lactones, such as helenalin (**136**), 11α,13-dihydrohelenalin (**138**), chamissonolid (**139**) and their ester derivatives. While these compounds affect various cellular processes, current data do not fully explain how sesquiterpene lactones exert their anti-inflammatory effects. We show here that helenalin (**136**), and, to a much lesser degree, 11α,13-dihydrohelenalin (**138**) and chamissonolid (**139**), inhibit activation of transcription factor NF-κB. This difference in efficacy, which correlates with the compounds' anti-inflammatory potency *in vivo*, may be explained by differences in structure and conformation. NF-κB, which resides in an inactive, cytoplasmic complex in unstimulated cells, is activated by phosphorylation and degradation of its inhibitory subunit, IκB. Helenalin inhibits NF-κB activation in response to four different stimuli in T-cells, B-cells and epithelial cells and abrogates κB-driven gene expression. This inhibition is selective, as the activity of four other transcription factors, Octamer transcription factor-1 (Oct-1), TATA -binding protein (TBP), Sp1 and Signal transducer and activator of transcription 5 (STAT 5) was not affected. We show that inhibition is not due to a direct modification of the active NF-κB heterodimer.

Rather, helenalin modifies the NF-κB/IκB complex, preventing the release of IκB. These data suggest a molecular mechanism for the anti-inflammatory effects of sesquiterpene lactones, which differs from that of other nonsteroidal anti-inflammatory drugs (NSAIDs), indomethacin and acetyl salicylic acid.

From the flower heads of *Arnica lonchophylla* ssp. *lonchophylla*, a variety of mono-, di- and tri-hydroxytriterpenes of the oleanane, ursane, lupane and dammarane types were isolated, and their structures were elucidated by EI and CI mass spectrometry, as well as extensive NMR spectroscopic analyses [42]. Most of the compounds are esterified at position 3 with lauric, myristic, palmitic and stearic acids. Several ester derivatives of known triterpenes represent new natural products. The triterpenetriols heliantriol F (**142**), arnitriol A (**143**) lupane-3β,16β,20-triol (**144**) and fouquierol (**145**), both isolated in the form of their C3-fatty acid ester derivatives, possess hydroxylation patterns that have not been described previously. In addition to small amounts of triterpenetriol esters, the lauric, myristic, palmitic and stearic acid esters of arnidiol, faradiol, maniladiol and calenduladiol were also identified in the flower heads of *A. montana* (Arnicae flos), *A. chamissonis* ssp. *foliosa* and *A. angustifolia* ssp. *attenuata*. A search for sesquiterpene lactones in the flowers of *A. lonchophylla* resulted in the identification of small amounts of helenalin (**136**) and 11α,13-dihydrohelenalin (**138**).

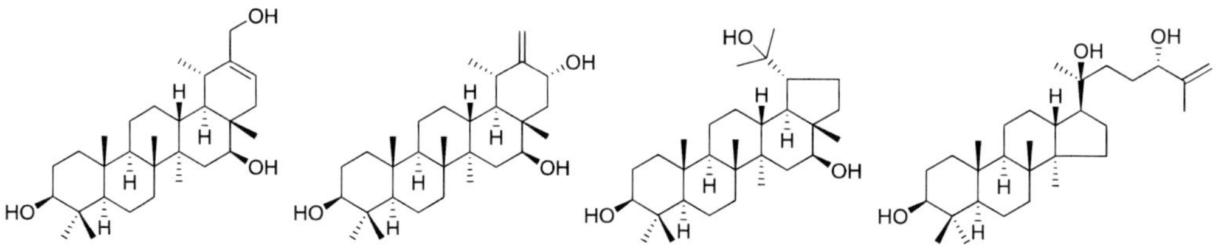

Figure 16. Structures of triterpenoids from flowerheads of Arnica lonchophylla.

Five flavonoid glycosides were identified from flowers of *A. montana*, four from *A. chamissonis* subsp. *foliosa* var. *incana* [43]. The structures were established based on total acid hydrolysis and spectral data as hispidulin 7-*O*-β-glucoside (**152**), isorhamnetin 3-*O*-β-glucoside (**112**), 3-*O*-β-D-glucopyranosids of spinacetin (**168**), 6-methoxykaempferol (**165**) and patuletin (**167**) and quercetin 3-*O*-(6"-*O*-acetyl)-β-D-glucopyranoside (**177**). The latter compound can serve as distinctive marker between these two *Arnica* species.

Two flavonol glycosides were isolated from flowers of *A. chamissonis* subsp. *foliosa* var. *incana* [44]. The structures were established based on acid hydrolysis and spectral data as the 7-β-glucosides of pectolinaringenin (**153**), hispidulin (**152**), jaceosidin (**155**), and eupafolin (**154**); isorhamnetin 3-β-glucoside (**113**); the 3-β-glucuronides of patuletin (**176**) and quercetin (**173**); and the 3-(6"-acetyl)-β-glucosides of kaempferol (**177**), 6-methyoxykaempferol (**180**), and patuletin (**181**). Five flavonoid glycosides were identified from the flowers of *A. Montana* [45]. The structures were established based on total acid hydrolysis and spectral data as the 3-*O*-β-D-glucuronides of quercetin (**173**), kaempferol (**172**), isorhamnetin (**174**), patuletin (**176**), and 6-methoxykaempferol (**175**).

Five flavonoid glycosides were identified from flowers or *A. montana* and 10 from *A. chamissonis foliosa incana* [46]. Structures were established based on acid hydrolysis and spectral data (UV, NMR and Mass spectrometry (MS)) as the 7-β-glucosides of pectolinarigenin (**153**), apigenin (**127**), and chrysoeriol (**148**); luteolin 3'-*O*-β-glucoside (**147**); the 3-β-glucuronides of kaempferol (**172**), isorhamnetin (**174**), and 6-methoxykaempferol (**175**); the 3,7-di-β-glucosides of quercetin (**165**) and patuletin (**171**); the 3-β-glucosides of betuletol (**167**) and quercetagetin 6,3',4'-trimethyl ether (**170**); and the 7-[6"-*O*-(2-methylbutyryl)] glucosides of luteolin (**149**) and eupafolin (**156**). The latter four are new natural compounds.

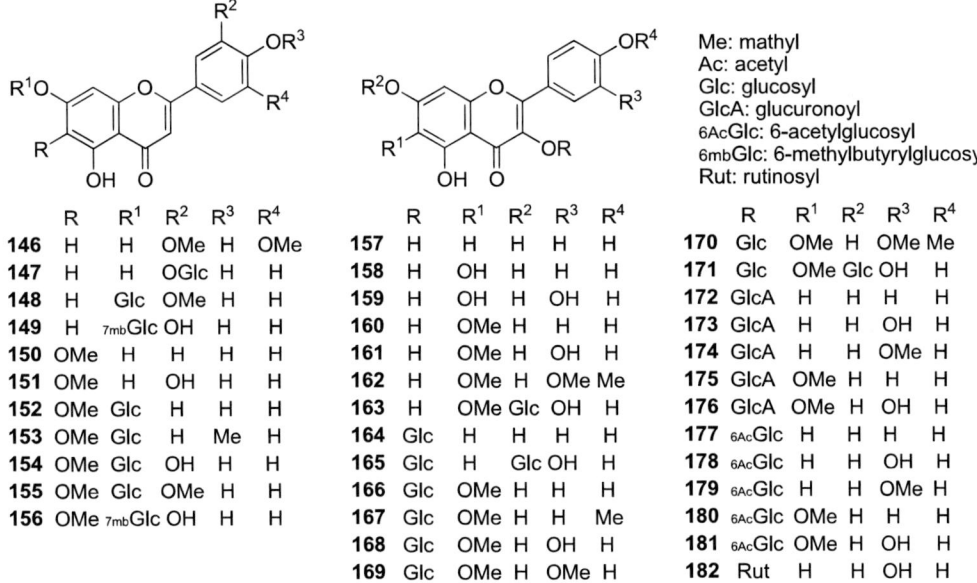

	R	R¹	R²	R³	R⁴
146	H	H	OMe	H	OMe
147	H	H	OGlc	H	H
148	H	Glc	OMe	H	H
149	H	7mbGlc	OH	H	H
150	OMe	H	H	H	H
151	OMe	H	OH	H	H
152	OMe	Glc	H	H	H
153	OMe	Glc	H	Me	H
154	OMe	Glc	OH	H	H
155	OMe	Glc	OMe	H	H
156	OMe	7mbGlc	OH	H	H

	R	R¹	R²	R³	R⁴
157	H	H	H	H	H
158	H	OH	H	H	H
159	H	OH	H	OH	H
160	H	OMe	H	H	H
161	H	OMe	H	OH	H
162	H	OMe	H	OMe	Me
163	H	OMe	Glc	OH	H
164	Glc	H	H	H	H
165	Glc	H	Glc	OH	H
166	Glc	OMe	H	H	H
167	Glc	OMe	H	H	Me
168	Glc	OMe	H	OH	H
169	Glc	OMe	H	OMe	H

	R	R¹	R²	R³	R⁴
170	Glc	OMe	H	OMe	Me
171	Glc	OMe	Glc	OH	H
172	GlcA	H	H	H	H
173	GlcA	H	H	OH	H
174	GlcA	H	H	OMe	H
175	GlcA	OMe	H	H	H
176	GlcA	OMe	H	OH	H
177	6AcGlc	H	H	H	H
178	6AcGlc	H	H	OH	H
179	6AcGlc	H	H	OMe	H
180	6AcGlc	OMe	H	H	H
181	6AcGlc	OMe	H	OH	H
182	Rut	H	H	OH	H

Me: mathyl
Ac: acetyl
Glc: glucosyl
GlcA: glucuronoyl
6AcGlc: 6-acetylglucosyl
6mbGlc: 6-methylbutyrylglucosyl
Rut: rutinosyl

Figure 17. Structures of flavonoids from Arnica flower.

The effects of the flavones apigenin (**44**), luteolin (**45**), hispidulin (**150**) and eupafolin (**151**), and of the flavonols kaempferol (**157**), quercetin (**111**), 6-methoxykaempferol (**160**) and patuletin (**161**) from *Arnica* spp. on the cytotoxicity of the sesquiterpene lactone helenalin (**136**) were studied in the human lung carcinoma cell line GLC$_4$ using microculture tetrazolium (MIT) assay [47]. Tumor cells were exposed to the test compounds for 2 h. Helenalin (**136**) concentrations around the control IC$_{50}$ value (0.5 µM) were combined with flavonoid concentrations ranging from 0.01 to 20 µM. At non-toxic concentrations of up to 10 µM, all flavonoids except kaempferol (**157**) significantly reduced helenalin-induced cytotoxicity. Hispidulin (**150**) and patuletin (**161**) displayed their modulating effects on helenalin-induced cytotoxicity in the broadest concentration range. The strongest effects were observed with 5 and 10 µM hispidulin (**150**), 0.05 µM quercetin (**111**), and 1 µM patuletin (**161**), increasing the IC$_{50}$ value of helenalin by approx. 40%. No dose-dependency was observed in the concentration range tested.

4.7. Tagetes

Tagetes is a genus of 56 species of annual and perennial mostly herbaceous plants in the sunflower family (Asteraceae). The genus is native to North and South America, but some species have become naturalized around the world. The florets of *T. erecta* are rich in the orange-yellow carotenoid lutein and are used as a food color in the European Union for foods such as pasta, vegetable oil, margarine, mayonnaise, salad dressing, baked goods, confectionery, dairy products, ice cream, yogurt, citrus juice and mustard. French Marigolds are commonly planted in butterfly gardens as a nectar source. Medicinally, various cultures have used infusions from dried leaves or florets. *T. erecta*, the Mexican marigold, also called Aztec marigold, is a species of the genus *Tagetes* native to Mexico and Central America. The Aztecs gathered the wild plant, as well as cultivating it, for medicinal, ceremonial and decorative purposes. This marigold may help protect certain crop plants from nematode pests when planted in fields. It is most effective against the nematode species *Pratylenchus penetrans*. Today, *T. erecta* is grown to extract lutein (**183**), a common yellow/orange food color. *Tagetes minuta*, also known as southern cone marigold, stinking roger or black mint, is a tall upright marigold plant from the genus *Tagetes*, with small flowers, native to the southern half of South America. It is used as a culinary herb in Peru, Ecuador, and parts of Chile and Bolivia. *T. lucida* is a half-hardy sub-shrub native to Mexico and Central America. It is eaten as an herb and is commonly used as a substitute for tarragon. Common names include Mexican marigold, pericon, Mexican mint marigold, Mexican tarragon, Spanish tarragon, and Texas tarragon. A pleasant anise-flavored tea, popular in Latin America, is brewed using the dried leaves and flowering tops. The petals are used as a condiment. A yellow dye can be obtained from the flowers.

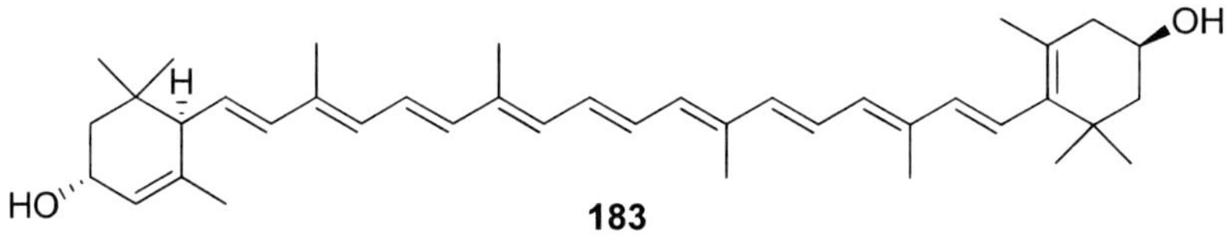

Figure 18. Structure of lutein (**183**) from French marigold.

The methanol extract of French marigold (*Tagetes patula*) florets (MEFMF) were found to inhibit acute and chronic inflammation in mice and rats [48]. MEFMF significantly suppressed hind-paw edema induced by γ-carrageenan in mice. Furthermore, MEFMF not only inhibited the hind-paw edema induced by various acute phlogogens, histamine, serotonin, bradykinin and prostaglandin E_1, but also suppressed the increase in vascular permeability by acetic acid, indicating that it primarily acts at the exudative stage of inflammation. In chronic inflammation, MEFMF did not inhibit the proliferation of granulation tissue when tested by the cotton pellet method; however, it was effective on the development of adjuvant arthritis in rats. Oral MEFMF inhibited acute and chronic inflammation in mice and rats.

The major components patuletin and patulitrin were isolated from French marigold (florets of *T. patula*). Patuletin and patulitrin were found to inhibit acute inflammation in mice [49]. Oral administration of patuletin (**161**) and patulitrin (**163**) significantly suppressed hind-paw edema induced by carrageenan and histamine, while topical application of patuletin and patulitrin significantly inhibited ear edema induced by TPA and arachidonic acid. Thus, oral and topical administration of patuletin (**161**) and patulitrin (**163**) inhibited acute inflammation in mice. These results suggest the anti-inflammatory efficacy of French marigold.

The antibacterial and antifungal activities of extracts from different parts of *T. patula* were evaluated in a single set of experiments [50]. In the preliminary assay, the methanol extract of the flower was found to possess antimicrobial activity against a number of bacteria with inhibition zone diameters ranging from 9 to 20 mm, the bioassay-guided fractionation of which led to the isolation of a flavonoid patuletin (**161**) in high yield as the active antibacterial principle with MIC value of 12.5 µg/disk against *Corynebacterium* spp., *Staphylococcus* spp., *Streptococcus* spp., and *Micrococcus luteus*. Its glucoside, patulitrin (**163**), was found to be weakly active, except against *Staphylococcus saprophyticus*, *Streptococcus fecalis* and *Streptococcus pyogenes* with inhibition zone diameters of 11, 16, and 12 mm, respectively. The cinnamate derivative of patuletin (**161**) showed antibacterial activity comparable with the parent flavonoid with a MIC value of 50 µg/disk against *Corynebacterium* spp., whereas benzoate derivative was found to be devoid of any activity; both the derivatives are new compounds. Moreover, the long chain alcohol tetracosanol ($C_{24}H_{49}OH$), which displayed antibacterial activity in the preliminary testing, was obtained in large quantites directly from the petroleum ether extract of the involucre of the flowers.

The search for more effective but non-toxic compounds with antioxidative potential led to bioassay-guided isolation studies on the extracts of *T. patula* [51]. The bioassay of *T. patula* flowers was carried out based on in vitro antioxidant activity using 1,1-diphenyl-2-picrylhydrazyl (DPPH) assay. A minor but proven plant constituent, methyl protocatechuate (**184**), was isolated by column chromatography, while patuletin and patulitrin were obtained in bulk by solvent partition of the methanol extract. Derivatization of patuletin into benzoyl, cinnamoyl and methyl was conducted in order to establish the structure activity relationship (SAR). Analgesic activity of patuletin (**161**) was evaluated using the acetic acid-induced writhing test and hot-plate test in mice. The toxicity of methanol extract and patuletin (**161**) were also determined. Polar extracts, fractions and phases demonstrated better antioxidant activity. The synthetic methyl protocatechuate (**184**) showed an IC_{50} value of 2.8 ± 0.2 µg/mL, whereas patuletin (IC_{50} = 4.3 ± 0.25 µg/mL) was comparable to quercetin (**111**) and rutin (**182**), but significantly better than patulitrin (IC_{50} = 10.17 ± 1.16 µg/mL). Toxicity test for the methanol extract and patuletin (**161**) did not elicit any behavioral changes or cause mortality in mice. patuletin (**161**) also demonstrated mild analgesic properties. These findings demonstrate that the plant polar extracts and fractions possess significant antioxidant property with a non-toxic effect. Methyl protocatechuate (**184**) is a genuine plant constituent of *T. patula*.

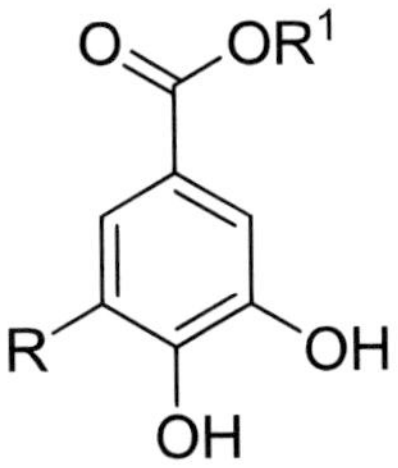

	R	R^1
184	H	Me
185	OH	H
186	OH	Me

Me: methyl

Figure 19. Structures of phenolics from *Tagetes* spp.

The influence of various solvents on the yield of polyphenols from defatted marigold residue, the antioxidant activity of the extracts and the composition of antioxidant compounds in the extracts were investigated [52]. The content of total phenolics and flavonoids in the extracts was significantly varied with different solvents (P<0.05) and the extract by ethyl alcohol (EtOH)/water (7:3, v/v) has the highest content of total phenolics and flavonoids, 62.33 mg gallic acid equivalents (GAE)/g and 97.00 mg rutin equivalent (RE)/g, respectively. The antioxidant activity of the extracts was evaluated by radical (2,2′-azino-bis-(3-ethylbenzothiazoline-6-sulfonic acid) (ABTS), DPPH scavenging and ferric reducing antioxidant power (FRAP) assays. The results of the correlation analysis showed that the antioxidant activity was well correlated with the content of total phenolics and flavonoids (R2>0.900). Antioxidant components in the extracts were identified by combined on-line HPLC–ABTS•+ post-column assay. Gallic acid (**185**), gallicin (**186**), quercetagetin (**159**), 6-hydroxykaempferol-*O*-glycoside, patuletin-*O*-glycoside and quercetin (**111**) were the dominant antioxidant compounds in the extracts, and quercetagetin was identified as having the strongest antioxidant capacity.

4.8. Sunflower

The sunflower (*Helianthus annuus*) is an annual plant native to the Americas. It possesses a large inflorescence (flowering head), and its name is derived from the flower's shape and image, which is often used to depict the sun. The heads consist of many individual flowers that mature into seeds, often in the hundreds, on a receptacle base. From the Americas, sunflower seeds were brought to Europe in the 16th century, where, along with sunflower oil, they became a widespread cooking ingredient. Although it was commonly accepted that the sunflower was first domesticated in what is now the Southeastern US, roughly 5000 years ago [53], there is evidence that it was first domesticated in Mexico [54] around 2600 B.C.E., as crops were found in Tabasco, Mexico, at the San Andre's dig site. Many indigenous American peoples used the sunflower as the symbol of their solar deity, including the Aztecs and the Otomi of Mexico and the Incas in South America. In 1510, early Spanish explorers encountered the sunflower in the Americas and took its seeds back to Europe.

A novel 3,4-*seco*-triterpene alcohol with a migrated euphane-type skeleton, called helianol (**12**), was isolated together with seven known triterpene alcohols from the nonsaponifiable lipid of the tubular flower extract of sunflower [55]. The structure was determined by extensive spectroscopic analysis.

Two new oleanane-type triterpene glycosides, named helianthosides 4 (**190**) and 5 (**191**), along with four known triterpene glycosides, helianthosides 1 (**187**), 2 (**188**), 3 (**189**), and B (**192**), were isolated from an *n*-butanol-soluble fraction of a methanol extract of sunflower petals [56]. Structures of the two new compounds were determined on the basis of spectroscopic and chemical methods. Upon evaluation of compounds 187~192 for inhibitory activity against TPA-induced inflammation (1.7 nmol/ear) in mice, all of the compounds tested exhibited marked anti-inflammatory activity, with ID_{50} values in the range 65-262 nmol/ear.

Fractionation of a petroleum ether extract of flower heads without seeds of *Helianthus annuus* led to the isolation of three diterpene acids: grandiflorolic (**193**), kaurenoic (**194**) and trachylobanoic (**195**) acids [57]. These compounds were studied for potential anti-inflammatory activity on the generation of inflammatory mediators in LPS-activated RAW 264.7 macrophages. At non-toxic concentrations, these compounds reduced, in a concentration-dependent manner, nitric oxide (NO), prostaglandin E_2 (PGE2) and tumor necrosis factor (TNF-α) production, as well as expression of inducible NO synthase (iNOS) and cyclooxygenase-2 (COX-2).

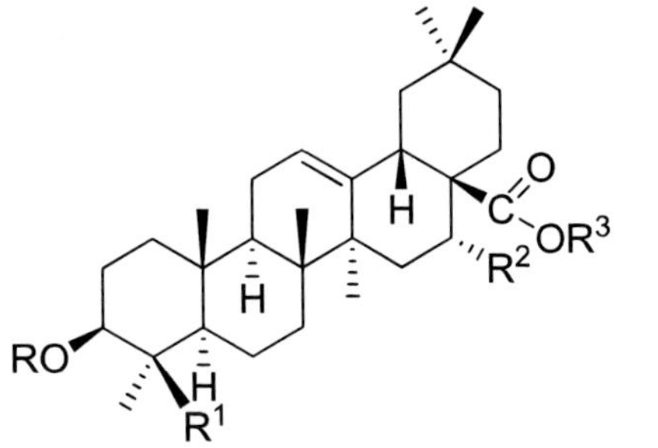

	R	R^1	R^2	R^3
187	S-4	Me	H	S-6
188	S-1	Me	OH	S-6
189	S-4	Me	OH	S-7
190	S-1	CH$_2$OH	OH	S-2
191	S-1	Me	OH	S-3
192	S-5	Me	OH	S-6

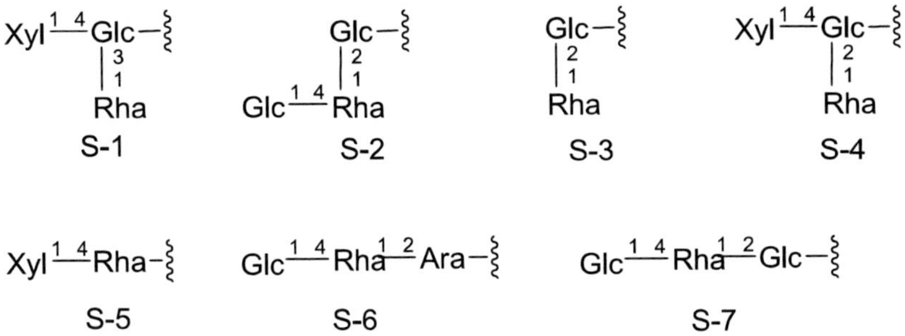

Me: methyl
Ara: arabinosyl
Glc: glucosyl
Rha: rhamnosyl
Xyl: xylosyl

Figure 20. Structures of saponins from sunflower.

All diterpenoids displayed significant in vivo anti-inflammatory activity and suppressed the TPA-mouse ear edema. In addition, inhibition of myeloperoxidase (MPO) activity, an index of cellular infiltration, was observed. In summary, our results suggest that the inhibition of the expression of NOS, COX-2 and the release of inflammatory cytokines, is responsible for the anti-inflammatory effects of the diterpenoids isolated from *H. annuus*, which likely contributes to the pharmacological action of sunflower.

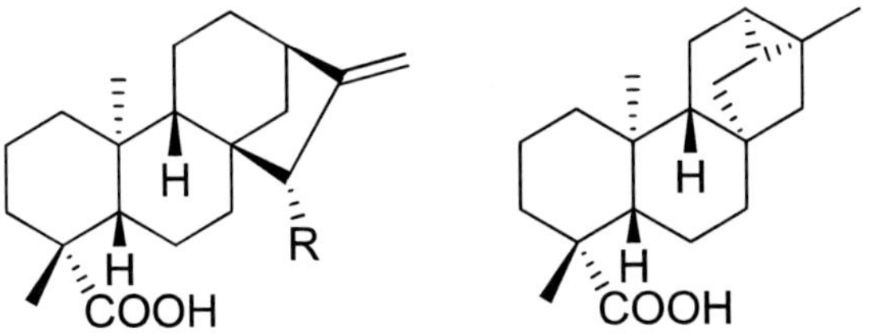

	R
193	H
194	OH

195

Figure 21. Structures of diterpenoids from sunflower.

4.9. Coltsfoot

Tussilago farfara, commonly known as coltsfoot, is a plant in the Asteraceae that has traditionally had medicinal uses. However, the discovery of toxic pyrrolizidine alkaloids in the plant has resulted in liver health concerns. The name "tussilago" itself means "cough suppressant". *Tussilago* is often found in colonies of dozens of plants. The flowers, which superficially resemble dandelions, appear in early spring before dandelions appear. Thus, the flowers appear on stems with no apparent leaves, and the leaves appear later and then wither and die during the season. Coltsfoot is also worked by the honey bee. Coltsfoot contains pyrrolizidine alkaloids, senecionine (**195**), integerrimine (**196**), seneciphylline (**197**), tussilogine (**198**), and isotussilogine (**199**) [58]. There are documented cases of coltsfoot tea causing severe liver problems in an infant, and in another case, an infant developed liver disease and died because the mother drank tea containing coltsfoot during her pregnancy [59,60]. In response, the German government banned the sale of coltsfoot. Clonal plants of coltsfoot free of pyrrolizidine alkaloids were then developed in Austria and Germany [61]. This has resulted in the development of the registered variety *T. farfara* which has no detectable levels of these alkaloids.

Figure 22. Structures of pyrrolizidine alkaloids from *Arnica* flowers.

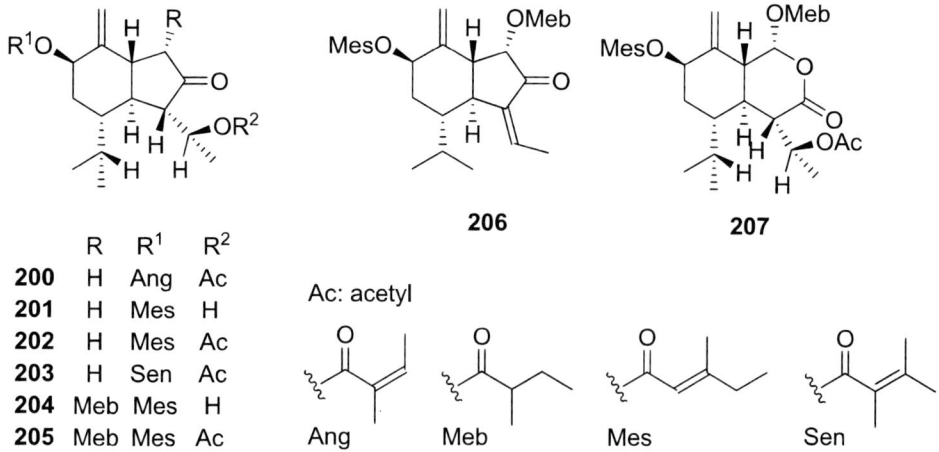

	R	R^1	R^2
200	H	Ang	Ac
201	H	Mes	H
202	H	Mes	Ac
203	H	Sen	Ac
204	Meb	Mes	H
205	Meb	Mes	Ac

Ac: acetyl

Ang   Meb   Mes   Sen

Fugure 23. Structure of sesquiterpenoids from flower of *Tussilago farfara*.

Four new notonipetranone-type sesquiterpenoids, 7β-(3-ethyl *cis*-crotonoyloxy)-14-hydroxy-notonipetranone (**201**), 14-acetoxy-7β-angeloyloxy-notonipetranone (**200**), 14-acetoxy-7β-senecioyloxy-notonipetranone (**203**), and tussilagolactone (**207**), were isolated together with 7β-(3-ethyl *cis*-crotonoyloxy)-14-hydroxy-1α-(2-methylbutyryloxy)-notonipetranone (**204**), and 7β-(3-ethyl *cis*-crotonoyloxy)-1α-(2-methylbutyryloxy)-3,14-dehydro-*Z*-notonipetranone (**206**) from the flower buds of *Tussilago farfara*. The structures of the new compounds were elucidated based on spectroscopic evidence.

Tussilagone (**202**), isolated from the flower of buds of *T. farfara*, is a sesquiterpenoid that is known to exert a variety of pharmacological activities. In the present study, we demonstrated that tussilagone (**202**) exerts anti-inflammatory activities in murine macrophages by inducing heme oxygenase-1 (HO-1) expression. Treatment of RAW264.7 cells with tussilagone (**202**)-induced HO-1 protein expression in a dose- and time-dependent manner without the induction of HO-1 mRNA expression. TSL-mediated HO-1 protein induction was not inhibited by treatment with actinomycin D, a transcriptional inhibitor, but by cycloheximide, a translational inhibitor. Moreover, MAPK inhibitors such as SB203580, SP600125 and U0126 did not block Tussilagone (**202**)-mediated HO-1 protein expression, thus suggesting that the Tussilagone (**202**)-mediated HO-1 induction may be regulated at the translational level. Consistent with the notion that HO-1 has anti-inflammatory properties, tussilagone (**202**) inhibited the production of NO, TNF-α and PGE2, as well as iNOS and COX-2 expression in LPS-stimulated RAW264.7 cells and murine peritoneal macrophages. Inhibition of HO-1 activity by treatment with ZnPP, a specific HO-1 inhibitor, abrogated the inhibitory effects of tussilagone (**202**) on the production of NO and PGE2 in LPS-stimulated RAW264.7 cells. Taken together, tussilagone (**202**) may be an effective HO-1 inducer that has anti-inflammatory effects, and a valuable compound for modulating inflammatory conditions.

CONCLUSION

The flowers of Asteraceae have long contributed to the health of human populations. We have provided an outline of the active components of these flowers, which have anti-inflammatory, cancer preventive, anti-bacterial and antitumor activities. For some of these, we also discussed the mechanisms by which they operate. In future studies, we believe that we can further elucidate the mechanisms of the active compounds of Asteraceae. The flowers of Asteraceae will continue to contribute to primary care in many countries around the world.

ACKNOWLEDGMENT

I have been studying natural bioactive products for approximately 40 years, and have compiled this chapter based on some of my research. My studies would have been impossible without the support of my wife and family, and this article is presented in gratitude of their support.

REFERENCES

[1]   Jeffrey, C. Compositae: Introduction with key to tribes. *In*: *Families and Genera of Vascular Plants,* vol. VIII, Flowering Plants, Eudicots, Asterales. Eds., Kadereit, J.W.; Jeffrey, C., Springer-Verlag, Berlin, 2007, pp 61–87.

[2]   Thompson, W.R.; Meinwald, J.; Aneshansley, D.; Eisner, T. Flavonols: pigments responsible for ultraviolet absorption in nectar guide of flower. *Science*, 1972; 177: 528–530.

[3]   Yasukawa, K.; Akihisa, T.; Inoue, Y.; Tamura, T.; Yamanouchi, S.; Takido, M. Inhibitory effect of the methanol extracts from Compositae plants on 12-*O*-tetradecanoylphorbol-13-acetate-induced inflammatory ear oedema in mice. *Phytother. Res.*, 1998, 12, 484–487.

[4]   Akihisa, T.; Yasukawa, K.; Oinuma, H.; Kasahara, Y.; Yamanouchi, S.; Takido, M.; Kumaki, K.; Tamura, T. Triterpene alcohols from the flowers of Asteraceae and their anti-inflammatory effects. *Phytochemistry*, 1996, 43, 1255–1260.

[5]   Yasukawa, K.; Akihisa, T.; Oinuma, H.; Kasahara, Y.; Kimura, Y.; Yamanouchi, S.; Kumaki, K.; Tamura, T.; Takido, M. Inihibitory effect of di- and trihydroxy triterpenes from the flowers of Asteraceae on 12-*O*-tetradecanoylphorbol-13-acetate-induced inflammation in mice. *Biol. Pharm. Bull.*, 1996, 19, 1329–1331.

[6]   Akihisa, T.; Franzblau, S.G.; Ukiya, M.; Okuda, H.; Zhang, F.; Yasukawa, K.; Suzuki, T.; Kimura, Y. Antitubercular activity of triterpenoids from Asteraceae flowers. *Biol. Pharm. Bull.*, 2005, 28, 158–160.

[7]   Rajic, A.; Akihisa, T.; Ukiya, M.; Yasukawa, K.; Sandeman, R.M.; Chandler, D.S.; Polya, G.M. Inhibition of trypsin and chymotrypsin by anti-inflammatory triterpenoids from asteraceae flowers. *Planta Med.*, 2001, 67, 599–604.

[8]   Rottenberg, A.; Zohary, D. The wild ancestry of the cultivated artichoke. *Genet. Res. Crop Evol.* 1996, 43, 53–58.

[9]   Ceccarelli, N.; Curadi, M.; Picciarelli, P.; Martelloni, L.; Sbrana, C.; Giovannetti, M. Globe artichoke as a functional food. *Mediterr. J. Nutr. Metab.*, 2010, 3, 197–201.

[10] Metwally, N.S.; Kholeif, T.E.; Ghanem, K.Z.; Farrag, A.R. Ammar, N.M.; Abdel-Hamid, A.H. The protective effects of fish oil and artichoke on hepatocellular carcinoma in rats. *Eur. Rev. Med. Pharmacol. Sci.*, 2011, 15, 1429–1444.

[11] Juzyszyn, Z.; Czerny, B.; Myśliwiec, Z.; Pawlik, A.; Droździk, M. The effect of artichoke (*Cynara scolymus* L.) extract on respiratory chain system activity in rat liver mitochondria. *Phytother Res.*, 2010, 24, S123-S128.

[12] Kirchhoff, R.; Beckers, C.; Kirchhoff, G.M.; Trinczek-Gärtner, H.; Petrowicz, O.; Reimann, H.J. Increase in choleresis by means of artichoke extract. *Phytomedicine*, 1994 , 1, 107-115.

[13] Qiang, Z.; Lee, S.O.; Ye, Z.; Wu, X.; Hendrich, S. Artichoke extract lowered plasma cholesterol and increased fecal bile acids in Golden Syrian hamsters. *Phytother Res.*, 2012, 26, 1048–1052.

[14] Englisch, W.; Beckers, C.; Unkauf, M.; Ruepp, M.; Zinserling, V. Efficacy of Artichoke dry extract in patients with hyperlipoproteinemia. *Arzneimittelforschung*, 2000, 50, 260–265.

[15] Gebhardt, R. Inhibition of Cholesterol Biosynthesis in Primary Cultured Rat Hepatocytes by Artichoke (*Cynara scolymus* L.) Extracts. *J. Pharmacol. Exp. Ther.*, 1998, 286, 1122–1128.

[16] Costabile, A.; Kolida, S.; Klinder, A.; Gietl, E.; Bäuerlein, M.; Frohberg, C.; Landschütze, V.; Gibson, G.R. A double-blind, placebo-controlled, cross-over study to establish the 'bifidogenic' effect of a very-long-chain inulin extracted from globe artichoke (*Cynara* scolymus) in healthy human subjects. *Br. J. Nutr.*, 2010, 104, 1007–1017.

[17] Holtmann, G.; Adam, B.; Haag, S.; Collet, W.; Grünewald, E.; Windeck, T. Efficacy of artichoke leaf extract in the treatment of patients with functional dyspepsia: a six-week placebo-controlled, double-blind, multicentre trial. *Aliment. Pharmacol. Ther.*, 2003, 18, 1099–1105.

[18] Bundy, R.; Walker, A.F.; Middleton, R.W.; Marakis, G.; Booth, J.C. Artichoke leaf extract reduces symptoms of irritable bowel syndrome and improves quality of life in otherwise healthy volunteers suffering from concomitant dyspepsia: a subset analysis. *J. Altern. Complement. Med.*, 2004, 10, 667–669.

[19] Walker, A.F.; Middleton, R.W.; Petrowicz, O. Artichoke leaf extract reduces symptoms of irritable bowel syndrome in a post-marketing surveillance study. *Phytother. Res.*, 2001, 15, 58–61.

[20] Yasukawa, K.; Matsubara, H.; Sano, Y. Inhibitory effect of the flowers of artichoke (*Cynara cardunculus*) on TPA-induced inflammation and tumor promotion in two-stage carcinogenesis in mouse skin. *J. Nat. Med.*, 2010, 64, 388–391.

[21] Zohary, D.; Hopf, M. Domestication of plants in the Old World, third edition, Oxford: University Press, 2000, p. 211.

[22] Kasahara, Y.; Yasukawa, K.; Kitanaka, S.; Khan, M.T.; Evans, F.J. Effect of methanol extract from flower petals of *Tagetes patula* L. on acute and chronic inflammation model. *Phytother. Res.*, 2002, 16, 217–22.

[23] Akihisa, T.; Oinuma, H.; Tamura, T.; Kasahara, Y.; Kumaki, K.; Yasukawa, K.; Takido, M. *erythro*-Hentriacontane-6,8-diol and 11 other alkane-6,8-diols from *Carthamus tinctorius*. *Phytochemistry*, 1994, 36, 105–108.

[24] Akihisa, T.; Nozaki, A.; Inoue, Y.; Yasukawa, K.; Kasahara, Y.; Motohashi, S.; Kumaki, K.; Tokutake, N.; Takido, M.; Tamura, T. Alkanediols from the flower petals of *Carthamus tinctorius*. *Phytochemistry*, 1997, 45, 725–728.

[25] Ukiya, M.; Akihisa, T.; Motohashi, S.; Yasukawa, K.; Kimura, Y.; Kasahara, Y.; Takido, M.; Tokutake, N. 6S,8R-Sterochemistry of the C_{27}- and C_{29}-alkane-6,8-diols isolated from three Asteraceae flowers. *Chem. Pharm. Bull. (Tokyo)*, 2000, 48, 1187–1189.

[26] Akihisa, T.; Inoue, Y.; Yasukawa, K.; Kasahara, Y.; Yamanouchi, S.; Kumaki, K.; Tamura, T. Widespread occurrence of *syn*-alkane-6,8-diols in the flowers of Asteraceae. *Phytochemistry*, 1998, 49, 1637–1640.

[27] Motohashi, S.; Akihisa, T.; Tamura, T.; Tokutake, N.; Takido, M.; Yasukawa, K. Alkane-6,8-diols: Inhibitor of tumor promotion in two-stage carcinogenesis in mouse skin. *J. Med. Chem.*, 1995, 38, 4155–4156.

[28] Miura, M.; Toriyama, M.; Kasahara, Y.; Akihisa, T.; Yasukawa, K.; Motohashi, S. Optically active alkane-6,8-diols as anti-tumor agents in mouse skin. *Phytochem. Lett.*, 2008, 1, 144–146.

[29] Yasukawa, K.; Akihisa, T.; Kasahara, Y.; Kaminaga, T.; Kanno, H.; Kumaki, K.; Tamura, T.; Takido, M. Inhibitory effect of alkane-6,8-diols, the components in safflower, on tumor promotion by 12-O-tetradecanoylphorbol-13-acetate in two-stage carcinogenesis in mouse skin. *Oncology*, 1996, 53, 133–136.

[30] Song, L.; Zhu, Y.; Jin, M.; Zang, B. Hydroxysafflor yellow a inhibits lipopolysaccharide-induced inflammatory signal transduction in human alveolar epithelial A549 cells. *Fitoterapia*, 2013, 84, 107–114.

[31] Ukiya, M.; Akihisa, T.; Yasukawa, K.; Tokuda, H.; Suzuki, T.; Kimura, Y. Anti-inflammatory, anti-tumor-promoting, and cytotoxic activities of constituents of marigold (*Calendula officinalis*) flowers. *J. Nat. Prod.*, 2006, 69, 1692–1696.

[32] White NJ. Assessment of the pharmacodynamic properties of antimalarial drugs in vivo. *Antimicrob Agents Chemother.* 1997, 41, 1413–1422.

[33] Islam, Md. N.; Jung, H. A.; Sohn, H. S.; Kim, H. M.; Choi, J. S. Potent α-glucosidase and protein tyrosine phosphatase 1B inhibitors from Artemisia capillaries. *Arch. Pharm. Res.*, 2013, 36, 542−552.

[34] Ukiya, M.; Akihisa, T.; Yasukawa, K.; Kasahara, Y.; Kimura, Y.; Koike, K.; Nikaido, T.; Takido, M. Constituents of Asteraceae plants. 2. Triterpenen diols, triols, and their 3-O-fatty acid esters from edible Chrysanthemum flower extract and their anti-inflammatory effects. *J. Agric. Food Chem.*, 2001, 49, 3187–3197.

[35] Ukiya, M.; Akihisa, T.; Tokuda, H.; Suzuki, H.; Mukainaka, T.; Ichiishi, E.; Yasukawa, K.; Kasahara, Y.; Nishino, H. Constituents of Asteraceae plants III. Anti-tumor promoting effects and cytotoxic activity against human cancer cell lines of triterpene diols and triols from edible chrysanthemum flowers. *Cancer Lett.*, 2002, 177, 7–12.

[36] Sugawara, T.; Igarashi, K. Identification of major flavonoids in petals of edible chrysanthemum flowers and their suppressive effect on carbon tetrachloride-induced liver injury in mice. *Food Sci. Technol. Res.*, 2009, 15, 499–506.

[37] Yasukawa, K.; Akihisa, T.; Oinuma, H.; Kaminaga, T., Kanno, H.; Kasahara, Y.; Tamura, T.; Kumaki, K.; Yamanouchi, S.; Takido, M. Inhibitory effect of taraxastane-type triterpenes on tumor promotion by 12-O-tetradecanoylphorbol-13-acetate in two-stage carcinogenesis in mouse skin. *Oncology*, 1996, 53, 341–344.

[38] Yasukawa, K.; Akihisa, T.; Kasahara, Y.; Ukiya, M.; Kumaki, K.; Tamura, T.; Yamanouchi, S.; Takido, M. Inhibitory effect of heliantriol C; a component of edible chrysanthemum, on tumor promotion by 12-O-tetradecanoylphorbol-13-acetate in two-stage carcinogenesis in mouse skin. *Phytomedicine*, 1998, 5, 215–218.

[39] Xie, Y.-Y.; Yuan, D.; Yang, J.-Y.; Wang, L.-H.; Wu, C.-F. Cytotoxic activity of flavonoids from the flowers of *Chrysanthemum morifolium* on human colon cancer Colon205 cells. *J. Asian Nat. Prod. Res.*, 2009, 11, 771–778.

[40] Widrig, R.; Suter, A.; Saller, R.; Melzer, J. Choosing between NSAID and arnica for topical treatment of hand osteoarthritis in a randomised, double-blind study. *Rheumatol. Int.*, 2007, 27, 585–591.

[41] Merfort, I.; Wendisch, D. Flavonoid glycosides from *Arnica montana* and *Arnica chamissonis*. *Planta Med.*, 1987, 53, 434−437.

[42] Merfort, I. Acetylated and other flavonoid glycosides from *Arnica chamissonis*. *Phytochemistry*, 1988, 27, 3281−3284.

[43] Merfort, I.; Wendisch, D. Flavonoid glucuronides from the flowers of *Arnica montana*. *Planta Med.*, 1988, 54, 247−250.

[44] Merfort, I.; Wendisch, D. New flavonoid glycosides from Arnicae flos DAB 9. *Planta Med.*, 1992, 58, 355–357.

[45] Woerdenbag, H.J.; Merfort, I.; Schmidt, T.J.; Passreiter, C.M.; Willuhn, G.; van Uden, W.; Pras, N.; Konings, A.W.T. Decreased helenalin-induced cytotoxicity by flavonoids from *Arnica* as studies in a human lung carcinoma cell line. *Phytomedicine*, 1995, 2, 127–132.

[46] Schmidt, T.J.; von Raison, J.; Willuhn, G. New triterpene esters from flowerheads of *Arnica lonchophylla. Planta Med.*, 2004, 70, 967–977.

[47] Lyss, G.; Schmidt, T.J.; Merfort, I.; Pahl, H.L. Helenalin, a anti-inflammatory sesquiterpene lactone from Arnica, selectively inhibits transcription factor NF-kappaB. *Biol. Chem.*, 1997, 378, 951–961.

[48] Kasahara, Y.; Yasukawa, K.; Kitanaka, S.; Khan, M.T.; Evans, F.J. Effect of methanol extract from flower petals of *Tagetes patula* L. on acute and chronic inflammation model. *Phytother. Res.*, 2002, 16, 217–222.

[49] Yasukawa, K.; Kasahara, Y. Effect of flavonoids from French marigold (florets of *Tagetes patula* L.) on acute inflammation *model. Int. J. Inflamm.*, submitted.

[50] Faizi, S.; Siddiqi, H.; Bano, S.; Naz, A.; Lubna; Mazhar, K.; Nasim, S.; Riaz, T.; Kamal, S.; Ahmad, A.; Khan, S.A. Antibacterial and antifungal activities of different parts of *Tagetes patula*: preparation of patuletin derivatives. *Pharm. Biol.*, 2008, 46, 309–320.

[51] Faizi, S.; Dar, A.; Siddiqi, H.; Naqvi, S.; Naz, A.; Bano, S.; Lubna. Bioassay-guided isolation of antioxidant agents with analgesic properties from flowers of *Tagetes patula. Pharm. Biol.*, 2011, 49, 516–525.

[52] Gong, Y.; Liu, X.; He, W.-H.; Xu, H.-G.; Yuan, F.; Gao, Y.-X. Investigation into the antioxidant activity and chemical composition of alcoholic extracts from defatted marigold (*Tagetes erecta* L.) residue. *Fitoterapia*, 2012, 83, 481–489.

[53] Blackman, B.K.; Scascitelli, M.; Kane, N.C.; Luton, H.H.; Rasmussen, D,A,; Bye, R,A,; Lentz, D.L.; Rieseberg, L.H. Sunflower domestication alleles support single domestication center in eastern North America. *Proc. Natl. Acad. Sci. USA.* 2011, 108, 14360–14365.

[54] Lentz, D.L.; Pohl, M.D.; Alvarado, J.L.; Tarighat, S.; Bye, R. Sunflower (*Helianthus annuus* L.) as a pre-Columbian domesticate in Mexico. *Proc. Natl. Acad. Sci. USA.*, 2008, 105, 6232–6237.

[55] Akihisa, T.; Oinuma, H.; Yasukawa, K.; Kasahara, Y.; Kimura, Y.; Takase, S.; Yamanouchi, S.; Takido, M.; Kumaki, K.; Tamura, T. Helianol [3,4-*seco*-19(10→9)*abeo*-8α,9β,10α-eupha-4,24-dien-3-ol], a novel triterpene alcohol from the tubular flowers of *Helianthus annuus* L. *Chem. Pharm. Bull. (Tokyo)*, 1996, 44, 1255–1257.

[56] Ukiya, M.; Akihisa, T.; Yasukawa, K.; Koike, K.; Takahashi, A.; Suzuki, T.; Kimura, Y. Triterpene glycosides from the flower petals of sunflower (*Helianthus annuus*) and their ant-inflammatory activity. *J. Nat. Prod.*, 2007, 70, 813–816.

[57] Díaz-Vicoedo, R.; Hortelano, S.; Girón, N.; Massó, J.M.; Rodriguez, B.; Villar, A.; de las Heras, B. Modulation of inflammatory responses by diterpene acids from *Helianthus annuus* L. *Biochem. Biophys. Res. Commun.*, 2008, 369, 761–766.

[58] Fu, P.P.; Yang, Y.C.; Xia, Q.; Chou, M.C.; Cui, Y.Y.; Lin G. Pyrrolizidine alkaloids-tumorigenic components in Chinese herbal medicines and dietary supplements. *J. Food Drug Anal.*, 2002, 10, 198−211.

[59] Sperl, W.; Stuppner, H.; Gassner, I.; Judmaier, W.; Dietze, O.; Vogel, W. Reversible hepatic veno-occlusive disease in an infant after consumption of pyrrolizidine-containing herbal tea. *Eur. J. Pediatr.*, 1995, 154, 112−116.

[60] Roulet, M.; Laurini, R.; Rivier, L.; Calame, A. Hepatic veno-occlusive disease in newborn infant of a woman drinking herbal tea. *J. Pediatr.*, 1988, 112, 433−436.

[61] Wawrosch, Ch.; Kopp, B.; Wiederfield, H. Permanent monitoring of pyrrolizidine alkaloid content in micropropagated *Tussilago farfara* L.: A tool to fulfill statutory demands for the quality of coltsfoot in Austria and Germany. *Acta Hortic.*, 2000, 530, 469−472.

[62] Kikuchi, M.; Suzuki, N. Studies on the constituents of *Tussilago farfara* L. II. Structures of new sesquiterpenoids isolated from the flower buds. *Chem. Pharm. Bull. (Tokyo)*, 1992, 40, 2753−2755.

[63] Hwangbo, C.; Lee, H.S.; Park, J.; Choe, J.; Lee, J.-H. The anti-inflammatory effect of tussilagone, from *Tussilago farfara*, is mediated by the induction of heme oxygenase-1 in murine macrophages. *Int. Immunopharmacol.*, 2009, 9, 1578−1584.

In: Flowers
Editors: Teodor Berntsen and Kaj Alsvik

ISBN: 978-1-62808-798-7
© 2013 Nova Science Publishers, Inc.

Chapter 2

ARACHIS SPECIES: CLASSIFICATION, PHYLOGENETIC STATUS AND USES AS ORNAMENTAL GROUNDCOVER

Chuan Tang Wang[1]*, Yue Yi Tang[1], Xiu Zhen Wang[1], Qi Wu[1], Zhen Yang[1], Qing Xuan Gong[1], Guo Sheng Song[1], Hua Yuan Gao[2], Wan Li Ni[3], Shu Tao Yu[4], Min Li[5], Lang Qian[5] and Tong Rong Yang[6]

[1.]Shandong Peanut Research Institute, Qingdao, China
[2]Institute of Peanut, Jilin Academy of Agricultural Sciences, Gongzhuling, China
[3]Institute of Crop Plants, Anhui Academy of Agricultural Sciences, Hefei, China
[4]Sandy Land Amelioration and Utilization Research Institute of Liaoning, Fuxin, China
[5]Dalian Academy of Agricultural Sciences, Dalian, China
[6]Wendeng Agricultural Bureau, Wendeng, China

ABSTRACT

The *Arachis* genus is divided into 9 taxonomic sections by Krapovicaks and Gregory (1994) based on the most important morphological characters. It is estimated that there are some 100 species within the genus. In the past, phylogenetic relationships were mainly inferred from morphological features, crossability, karyotype analysis and protein markers; in recent years, however, newer DNA markers and *in situ* hybridization techniques have been used in systematic studies of the genus, resulting in findings somewhat different from previously reported postulations. Globally, four wild peanut species have been used as ornamental groundcover. In Florida, USA, rhizoma perennial peanut, *A. glabrata*, is considered a good sod alternative in the urban landscape. In North Queensland, Australia, *A. pintoi* is used for bank stabilization and as ornamental groundcover in shaded and erosion-prone regions; in South China, the species has been extensively planted in orchards and gardens. *A. duranensis* is accepted as a good ornamental groundcover in South China; it also works well when grown in a hanging pot.

* Author for correspondence. chinapeanut@126.com.

A. repens is widely utilized as an ornamental both along roadsides and in home yards in South America. Up to now, most of the studies related to wild peanut species have been conducted by peanut breeders, rather than workers on flowers and turf science, with aims to enhance the productivity and stress resistances of *A. hypogaea* more often and in some cases to transfer desirable quality traits from wild peanut species. Research on utilization of wild *Arachis* species in breeding ornamental groundcover cultivars should be further strengthened.

Keywords: Peanut, *Arachis*, taxonomy, phylogenetic relationship, ornamental groundcover

1. INTRODUCTION

The *Arachis* genus is believed to originate in South America, in southwestern Brazil or northeast Paraguay (Simpson, 1995). Species of the genus are known to be geocarpic. Subterranean fruits (pods) are placed underground by the pegs. Though the pegs are not unique in plant taxa, very few similar structures have been found in the plant kingdom (Simpson, 1995).

Grown in most tropical, subtropical and temperate regions, the cultivated peanut or groundnut, *Arachis hypogaea* L., also known as common peanut, is a species of economic importance within the genus. It is considered as a main oilseed cash crop in developing countries, and a rich source of protein for human consumption in the developed world. Peanut can be eaten raw, roasted, boiled, or as a paste called peanut butter (Simpson, 1995). Peanut vines are used for animal feed. Shells can be used as fuel or as growing medium for mushrooms. Other uses of peanut shells include but not limited to production of activated carbon, particle boards, dietary fiber, a medicine (脉通灵) for lowering blood pressure and cholesterol level, sanitized odor-absorbing cat litter, and a substitute for phenolic resins, which are widely used in adhesives and for molding and casting, bonding, laminating and surface coating (Woodroof, 1983). Pigments from peanut seed coats are used in the manufacture of soy sauce. Seed coats and leaflets are thought to be effective in the treatment of thrombopenia or insomnia, respectively. Peanut kernels are well-known Chinese traditional medicine with multiple functions (Wan, 2003). In a time of increasing demand for energy supply and shortage of fossil fuels, interest in peanut as a source of renewable energy is growing (Wang et al., 2012). Cheap hot peanut meal can be processed into high value-added textile raw materials. Procedures for spinning peanut protein have been reported (Ahmed and Fletcher, 1977; Fletcher and Ahmed, 1977; Woodroof, 1983; Guan et al., 2006; Xing, 2005).

It is noteworthy that *A. hypogaea* L. is only a member of the *Arachis* genus. Several wild peanut species have been utilized in the genetic improvement of the cultivated peanut, and much progress has thus far been achieved (Wang et al., 2013a; Wang et al., 2013b). Forage use aspects of wild peanut species were summarized in the monograph entitled *Biology and agronomy of forage Arachis*, a CIAT publication (Kerridge and Hardy, 1994), by Cook and Crosthwaite (1994) in *The groundnut crop: a scientific basis for improvement*, and more recently by Wang et al. (2011a) in the book *Peanut genetic and breeding science in China* (Yu et al., 2011). With the development of economy, more and more people are getting to know the importance of human health, environment protection and landscaping. As a result, *Arachis* species, as an important and competitive ornamental groundcover with nitrogen fixation ability, beautiful flowers as well as inherited multiple and high resistances to harsh environment frequently encountered in their native habitats, are preferred by gardeners, city designers and ordinary people. But unfortunately, at present, there is no review regarding the global status of wild peanut species as ornamental groundcover. In addition, to make better use of wild peanut in landscaping, understanding their taxonomy and phylogenetic relationship is considered as a must. All those are included in the topics of this chapter.

2. TAXONOMY STATUS OF THE *ARACHIS* GENUS

Estimated to consist of about 100 species, the *Arachis* genus is apportioned to 9 sections (Krapovicaks and Gregory, 1994) on the basis of the most important morphological features, including flowers, pegs, fruits, rhizomatous and stoloniferous stems, root systems, hypocotyl and leaflet numbers (Simpson, 1995; Stalker and Simpson, 1995) (Table 1). Table 2 is a list of GRIN database records of *Arachis* species/subspecies/varieties, inclusive of synonyms, and their sectional affiliations.

Table 1. Key for identifying individual sections in the *Arachis* genus

A. Leaves trifoliolate. Hypocotyl tuberiform. Plants erect.
Flowers and fruits grouped around the collar of the plant.
Pegs horizontal, superficial and very long.

I. Trierectoides

A. Leaves tetrafoliolate. Hypocotyl cylindrical.

  B. Plants without rhizomes.

   C. Fruits with 2-3 articles. Branches decumbent.
Flowers and fruits along length of branches. Standard with
reddish lines on both sides. Cotyledons with veins on
upper surface very sunken.

IV Triseminatae

    C'. Fruits with 2 articles. Cotyledons with smooth upper
surface.
     D. Standard with red lines on the back side or on both
sides. Branches procumbent.
      E. Plants perennial. Roots with thickenings. Standard
with red lines only on the back side. All flowers normal,
with expanded corolla.

III. Extranervosae

      E'. Plants annual. Roots not thickened. Standard with
red lines on the back side or on both sides. Flowers
dimorphic: normal and open, or very small with corolla not
exceeding the calyx.

V. Heteranthae

     D'. Standard with red lines on the front side.

Table 1. (Continued)

F. Plants erect or decumbent. Flowers densely grouped around the collar of the plant; those flowers normally produce fruits. Towards the base of the branches, only buried flowers produce fruits. Roots with swollen branches (except in *A. stenophylla* and *A. paraguariensis*).	II. *Erectoides*
F'. Branches procumbent. Collars of the plant without flowers; inflorescences and fruits along the length of branches. In *A. appressipilla* (Section *Procumbentes*) the branches are decumbent, but the flowers are not grouped around the collar of the plant. G. Stems with roots at the nodes.	VI. *Caulorhizae*
G'. Stems without roots at the nodes, sometimes at the underground basal internodes. H. Pegs horizontal, very long and superficial.	VII. *Procumbentes*
H'. Pegs almost vertical.	IX. *Arachis*
B'. Plants rhizomatous.	VIII. *Rhizomatosae*

After Krapovickas and Gregory (1994) [translated by Williams and Simpson (2007)] and Stalker and Simpson (1995). © Instituto de Botánica del Nordeste, Argentina.

Table 2. GRIN database records of species/subspecies/botanical varieties within the *Arachis* genus, including synonyms, and their sectional affiliations

Species/subspecies/varieties	Chromosome number (2n)	Sections
1. Arachis appressipila **Krapov. & W. C. Greg.**	20	*Procumbentes*
2. Arachis archeri **Krapov. & W. C. Greg.**	20	*Erectoides*
3. Arachis batizocoi **Krapov. & W. C. Greg.**	20	*Arachis*
4. Arachis ×batizogaea **Krapov. & A. Fernandez**	-	
5. Arachis benensis **Krapov. et al.**	20	*Arachis*
6. Arachis benthamii **Handro**	20	*Erectoides*
7. Arachis brevipetiolata **Krapov. & W. C. Greg.**	n/d	*Erectoides*
8. Arachis burchellii **Krapov. & W. C. Greg.**	20	*Extranervosae*
9. Arachis burkartii **Handro**	20	*Rhizomatosae*
10. Arachis cardenasii **Krapov. & W. C. Greg.**	20	*Arachis*

Species/subspecies/varieties	Chromosome number (2n)	Sections
11. Arachis chacoense Krapov. & W. C. Greg., nom. nud. (=*Arachis diogoi* Hoehne)	20	*Arachis*
12. Arachis chiquitana Krapov. et al.	20	*Procumbentes*
13. Arachis correntina (Burkart) Krapov. & W. C. Greg.	20	*Arachis*
14. Arachis cruziana Krapov. et al.	20	*Arachis*
15. Arachis cryptopotamica Krapov. & W. C. Greg.	20	*Erectoides*
16. Arachis dardani Krapov. & W. C. Greg.	20	*Heteranthae*
17. Arachis decora Krapov. et al.	18	*Arachis*
18. Arachis diogoi Hoehne	20	*Arachis*
19. Arachis douradiana Krapov. & W. C. Greg.	20	*Erectoides*
20. Arachis duranensis Krapov. & W. C. Greg.	20	*Arachis*
21. Arachis fruticosa Retz. (=*Stylosanthes fruticosa* (Retz.) Alston)		na
22. Arachis giacomettii Krapov. et al.	20	*Heteranthae*
23. Arachis glabrata Benth.	40	
24. Arachis glabrata var. *glabrata*	40	*Rhizomatosae*
25. Arachis glabrata var. *hagenbeckii* (Harms) F. J. Herm.	40	*Rhizomatosae*
26. Arachis glandulifera Stalker	20	*Arachis*
27. Arachis gracilis Krapov. & W. C. Greg.	20	*Erectoides*
28. Arachis gregoryi C. E. Simpson et al.	20	*Arachis*
29. Arachis guaranitica Chodat & Hassl.	20	*Trierectoides*
30. Arachis hagenbeckii Harms (=*Arachis glabrata* var. *hagenbeckii* (Harms) F. J. Herm.)	40	*Rhizomatosae*
31. Arachis hassleri Krapov. et al.	**20**	***Procumbentes***
32. Arachis hatschbachii Krapov. & W. C. Greg.	**20**	***Erectoides***
33. Arachis helodes Mart. ex Krapov. & Rigoni	**20**	***Arachis***
34. Arachis hermannii Krapov. & W. C. Greg.	**20**	***Erectoides***
35. Arachis herzogii Krapov. et al.	**20**	***Arachis***
36. Arachis hoehnei Krapov. & W. C. Greg.	**20**	***Arachis***
37. Arachis hybr.		-
38. Arachis hypogaea L.	**40**	***Arachis***
39. Arachis hypogaea var. *aequatoriana* Krapov. & W. C. Greg.	**40**	***Arachis***
40. Arachis hypogaea subsp. *fastigiata* Waldron	**40**	***Arachis***

Table 2. (Continued)

Species/subspecies/varieties	Chromosome number (2n)	Sections
41. *Arachis hypogaea var. fastigiata* (Waldron) **Krapov. & W. C. Greg.**	40	*Arachis*
42. *Arachis hypogaea var. hirsuta* **J. Kohler**	40	*Arachis*
43. *Arachis hypogaea subsp. hypogaea*	40	*Arachis*
44. *Arachis hypogaea var. hypogaea*	40	*Arachis*
45. *Arachis hypogaea f. nambyquarae* (Hoehne) F. J. Herm. (=*Arachis hypogaea var. hypogaea*)	40	*Arachis*
46. *Arachis hypogaea var. nambyquarae* (Hoehne) Burkart (=*Arachis hypogaea var. hypogaea*)	40	*Arachis*
47. *Arachis hypogaea var. peruviana* **Krapov. & W. C. Greg.**	40	*Arachis*
48. *Arachis hypogaea subsp. sylvestris* A. Chev. (=*Arachis sylvestris* (A. Chev.) A. Chev.)	40	*Heteranthae*
49. *Arachis hypogaea var. vulgaris* **Harz**	40	*Arachis*
50. *Arachis interrupta* **Valls & C. E. Simpson**	20	*Heteranthae*
51. *Arachis ipaensis* **Krapov. & W. C. Greg.**	20	*Arachis*
52. *Arachis kempff-mercadoi* **Krapov. et al.**	20	*Arachis*
53. *Arachis krapovickasii* **C. E. Simpson et al.**	20	*Arachis*
54. *Arachis kretschmeri* **Krapov. & W. C. Greg.**	20	*Procumbentes*
55. *Arachis kuhlmannii* **Krapov. & W. C. Greg.**	20	*Arachis*
56. *Arachis lignosa* (Chodat & Hassl.) **Krapov. & W. C. Greg.**	20	*Procumbentes*
57. *Arachis linearifolia* **Valls et al.**	20	*Arachis*
58. *Arachis lutescens* **Krapov. & Rigoni**	20	*Extranervosae*
59. *Arachis macedoi* **Krapov. & W. C. Greg.**	20	*Extranervosae*
60. *Arachis magna* **Krapov. et al.**	20	*Arachis*
61. *Arachis major* **Krapov. & W. C. Greg.**	20	*Erectoides*
62. *Arachis marginata* **Gardner**	n/d	*Extranervosae*
63. *Arachis martii* **Handro**	n/d	*Erectoides*
64. *Arachis matiensis* **Krapov. et al.**	20	*Procumbentes*
65. *Arachis microsperma* **Krapov. et al.**	20	*Arachis*
66. *Arachis monticola* **Krapov. & Rigoni**	40	*Arachis*
67. *Arachis nambyquarae* Hoehne (=*Arachis hypogaea var. hypogaea*)	40	*Arachis*
68. *Arachis nitida* **Valls et al.**	40	*Rhizomatosae*

Species/subspecies/varieties	Chromosome number (2n)	Sections
69. *Arachis oteroi* **Krapov. & W. C. Greg.**	20	*Erectoides*
70. *Arachis palustris* **Krapov. et al.**	18	*Arachis*
71. *Arachis paraguariensis* **Chodat & Hassl.**	20	*Erectoides*
72. *Arachis paraguariensis* **subsp.** *capibarensis* **Krapov. & W. C. Greg.**	20	*Erectoides*
73. *Arachis paraguariensis* **subsp.** *paraguariensis*	20	*Erectoides*
74. *Arachis pflugeae* **C. E. Simpson et al.**	20	*Procumbentes*
75. *Arachis pietrarellii* **Krapov. & W. C. Greg.**	20	*Extranervosae*
76. *Arachis pintoi* **Krapov. & W. C. Greg.**	20	*Caulorrhizae*
77. *Arachis porphyrocalyx* **Valls & C. E. Simpson**	18	*Erectoides*
78. *Arachis praecox* **Krapov. et al.**	18	*Arachis*
79. *Arachis prostrata* **Benth.**	20	*Extranervosae*
80. *Arachis pseudovillosa* **(Chodat & Hassl.) Krapov. & W. C. Greg.**	40	*Rhizomatosae*
81. *Arachis pusilla* **Benth.**	20	*Heteranthae*
82. *Arachis rasteiro* A. Chev. (=*Arachis hypogaea* **subsp.** *hypogaea*)	40	*Arachis*
83. *Arachis repens* **Handro**	20	*Caulorrhizae*
84. *Arachis retusa* **Krapov. et al.**	20	*Extranervosae*
85. *Arachis rigonii* **Krapov. & W. C. Greg.**	20	*Procumbentes*
86. *Arachis schininii* **Krapov. et al.**	20	*Arachis*
87. *Arachis seridoensis* **Valls et al.**	20	*Heteranthae*
88. *Arachis setinervosa* **Krapov. & W. C. Greg.**	n/d	*Extranervosae*
89. *Arachis simpsonii* **Krapov. & W. C. Greg.**	20	*Arachis*
90. *Arachis spp.*	n/a	
91. *Arachis stenophylla* **Krapov. & W. C. Greg.**	20	*Erectoides*
92. *Arachis stenosperma* **Krapov. & W. C. Greg.**	20	*Arachis*
93. *Arachis subcoriacea* **Krapov. & W. C. Greg.**	20	*Procumbentes*
94. *Arachis submarginata* **Valls et al.**	20	*Extranervosae*
95. *Arachis sylvestris* **(A. Chev.) A. Chev.**	20	*Heteranthae*
96. *Arachis trinitensis* **Krapov. & W. C. Greg.**	20	*Arachis*
97. *Arachis triseminata* **Krapov. & W. C. Greg.**	20	*Triseminatae*
98. *Arachis tuberosa* **Benth.**	20	*Trierectoides*

Table 2. (Continued)

Species/subspecies/varieties	Chromosome number (2n)	Sections
99. Arachis valida **Krapov. & W. C. Greg.**	20	*Arachis*
100. Arachis vallsii **Krapov. & W. C. Greg.**	20	*Arachis*
101. Arachis villosa **Benth.**	20	*Arachis*
102. Arachis villosa var. *correntina* Burkart (=*Arachis correntina* **(Burkart) Krapov. & W. C. Greg.**)	20	*Arachis*
103. Arachis villosulicarpa **Hoehne**	20	*Extranervosae*
104. Arachis williamsii **Krapov. & W. C. Greg.**	20	*Arachis*

Source: USDA, ARS, National Genetic Resources Program. Species records in the GRIN database (http://www.ars-grin.gov/cgi-bin/npgs/html/tax_search.pl) (accessed May 19, 2013) and Lavia et al. (2008). Valid names are in bold. *Arachis dardani* was misspelled as *Arachis dardanoi* in GRIN the database.
n/a=not applicable.
n/d=no data.

3. PHYLOGENETIC RELATIONSHIP OF THE 9 SECTIONS OF THE *ARACHIS* GENUS

3.1. Genomes of the *Arachis* Genus

Husted (1933, 1936) first noted two distinctive pairs of chromosomes in *A. hypogaea*: a conspicuously small pair with medium centromere ("A" chromosomes), and a pair with unusually long secondary constriction and subterminal centromere ("B" chromosomes). Within Section *Arachis*, some diploid wild species were found to have "A" chromosomes, and several other ones lacked such a pair. Smartt et al. (1978) first proposed the concept of A and B genomes. *A. hypogaea* is considered as an AABB-genome species. In addition to A and B genomes, a distinctive D genome was identified by Stalker (1991) in a Section *Arachis* species, *A. glandulifera* Stalker, which had a highly asymmetrical karyotype with 6 subtelocentric or submetacentric chromosome pairs, in contrast to the A- and B-genome species mainly composed of metacentric chromosomes (Moretzsohn et al., 2013). Recently, based on FISH mapping of rDNA and heterochromatin detection using 11 non-A genome (or B-genome *sensu lato*) species, Robledo and Seijo (2010) propose segregating the non-A genome taxa into 3 genomes: B *sensu stricto* (s.s.), F and K. The B genome s.s., is deprived of centromeric heterochromatin and is homologous to one of the *A. hypogaea* complements (Robledo and Seijo, 2010). The other two genomes have centromeric bands on most of the chromosomes, but differ in the amount and distribution of heterochromatin: the K genome is characterized by karyotypes with large centromeric bands, and F genome has karyotypes with tiny centromeric bands in seven or eight chromosome pairs (Robledo and Seijo, 2010). Five former B-genome species change their genomic affiliations (Robledo and Seijo, 2010). Updated information on the genomes in the genus is summarized in Table 3.

Studies have indicated the existence of intrasectional and intersectional homologies. Within the *Arachis* section, when *A. batizocoi* or *A. glandulifera* is crossed with other species, though the hybrids are sterile, 6-8 chromosomes consistently pair during meiosis in these hybrids (Stalker, 1981). Stalker (1981) studied meiotic behavior of the hybrids between Section *Arachis* diploid species and an amphidiploid of Section *Erectoides* and found a high frequency of bivalents. Likewise, the hybrids between Section *Erectoides* and Section *Rhizomatosae* species also showed chromosomal homologies. Based on crossablity data, Smartt and Stalker (1982) proposed that tetraploid *Rhizomatosae* species have one genome in common with section *Erectoides*, and one closely related to section *Arachis* (Stalker and Simpson, 1995).

Table 3. Genomes in the *Arachis* genus

Genomes	Remarks
A1	Section *Arachis*, annual types
A2	Section *Arachis*, perennial types
B or B *sensu lato*	Section *Arachis*, species without "A" chromosomes (non-A genome species), *A. ipaensis, A. gregoryi, A. valida, A. williamsii, A. magna, A. batizocoi, A. cruziana A. krapovickasii, A. benensis, A. trinitensis* and an unnamed species (*A.* sp.)
B s.s	Section *Arachis*, *A. ipaensis, A. gregoryi, A. valida, A. williamsii, A. magna* and an unnamed species (*A.* sp.)
K	Section *Arachis*, *A. batizocoi, A. cruziana* and *A. krapovickasii*
F	Section *Arachis*, *A. benensis* and *A. trinitensis*
D	Section *Arachis*, *A. glandulifera*
H	Section *Heteranthae*
T	Section *Triseminatae*
C	Section *Caulorrhizae*
Ex	Section *Extranervosae*
E	Section *Erectoides*
Te	Section *Trierectoides*
R1	Section *Rhizomatosae*, Series *Prorhizomatosae*
R2	Section *Rhizomatosae*, Series *Rhizomatosae*
P	Section *Procumbensae*

After Simpson (1995), Smartt (1990), Singh and Simpson (1994) and Robledo and Seijo (2010).

3.2. Species with 2n=2x=18

Four wild peanut species, *A. decora, A. palustris, A. praecox* and *A. porphyrocalyx* were found to have 18 chromosomes only. The last species is a member of Section *Erectoides*. The first three belong to Section *Arachis*, having a common karyotype formulae (16m+2sm) with the same chromosome type 3 and lacking the A chromosome pair (Lavia et al., 2008). Though morphologically similar to A-genome species (Lavia et al., 2008), in the phylogenetic trees constructed by Wang et al. (2011b), these 3 species along with 1 D-genome species (*A.*

glandulifera) and 6 B-genome species together constitute a less advanced subgroup as compared with the subgroup mainly consisting of A- and AABB- genome species.

The origin of the species with x=9 is still a mystery unsolved. The average DNA content (2C) of the diploid species with x=10 and with x=9 from Section *Arachis* is 5.13 pg and 3.67 pg, respectively (Lavia et al., 2008). The great difference is unlikely caused by loss of a unique chromosome pair (Lavia et al., 2008).

3.3. Polyploidization

While most species in the genus *Arachis* are diploid (2n=2x-20), some species from Sections *Arachis* and *Rhizomatosae* are tetraploids (2n=2x=40). It is believed that tetraploids are more advanced and that polyploidization has evolved independently in the two sections (Stalker and Simpson, 1995). The cultivated peanut, *A. hypogae*a L., is believed to be an allotetraploid resulting from chromosome doubling of the hybrid between an A-genome and a B-genome species. Though a few quadrivalent have been observed in some accession of *A. hypogaea* L., consistent quadrivalent associations were only observed in tetraploid species of Section *Rhizomatosae*, supporting their possible autotetraploid nature (Singh, 1985; Singh and Simpson, 1994).

According to Lavia et al. (2008), it is still unclear if different subspecies and botanical varieties of the cultivated peanut are of simple or multiple origins and if introgression occurred during domestication and dispersion of the ancestral tetraploid. However, most results seem to support a single orgin. Grabiele et al. (2012) sequenced the non-coding cpDNA regions and non-transcriped spacer of nuclear 5S DNA, and the results strongly support that 6 botanical varieties of the cultivated peanut were of single genetic origin and that *A. monticola* is the immediate tetraploid ancestor from which the peanut cultigen has arisen upon domestication. Evidence from restriction fragment length polymorphism (RFLP), double GISH (genomic *in situ* hybridization), *trn* T-F region of the plastid genomes, intron sequences and re-synthesis of the allopolyploid supports the hypothesis that *A. duranensis* (A-genome donor) and *A. ipaensis* (B-genome donor) are the ancestral progenitors of the cultivated peanut (Lavia et al., 2008; Moretzsohn et al., 2013). Base on crossability studies, ICRISAT (International Crops Research Institute for the Semi-Arid Tropics) scientists, however, proposed *A. hoehnei* instead of *A. ipaensis* as the possible B-genome donor (Dr Nalini Mallikarjuna, personal communication).

RAPD (randomly amplified polymorphic DNA) analysis studies conducted by Nelson et al. (2006) revealed high levels of genetic diversity in the 4 populations of *A. glabrata* and genetic difference between *A. glabrata* and *A. pseudovillosa*, supporting the multiple origins of tetraploids in the Section *Rhizomatosae*. Genetic variation within and among species of Section *Rhizomatosae* was evaluated by Nóbile et al. (2004) using RAPD analysis. The diploid and tetraploid species are grouped quite separately, indicating that the tetraploids did not originate from the diploid species studied. This is also true in the phylogenteic trees constructed by Wang et al. (2011b).

3.4. Phylogeny

3.4.1. Intuition Inference Based on Morphology, Crossability and Geography

Species of Section *Trierectoides* (formerly Series *Trifoliolatae* of Section *Erectoides*) have trifoliolate leaves and once were considered most primitive. However, Singh and Simpson (1994) noticed the occurrence of *A. tuberosa* (a member of Section *Trifoliolatae*) populations with tetrafoliolate leaves, and therefore pointed out that the trifoliolate form may just be a simple genetic mutant. In their opinion, Section *Extranervosae* is very primitive: species from this section have unique pollen type (syncolpated pollen), and have developed strong genetic barriers, only crossable with species of Section *Heteranthae* (formerly Section *Ambinervosae*) (Singh and Simpson, 1994).

On the basis of information from morphology, crossability and geography, Krapovickas and Gregory (1994), in their most recent monograph on taxonomy of the *Arachis* genus, stated that, *Triseminatae, Trierectoides, Erectoides, Extranervosae* and *Heteranthae* are the presumably older sections, and the above-mentioned sections except for section *Erectoides* are much more isolated from the rest sections and from each other than those taken to be of more recent origin, viz., Sections *Procumbentes, Caulorrhizae, Rhizomatosae* and *Arachis* (Krapovickas and Gregory, 1994).

3.4.2. Molecular Phylogeny

Singh et al. (1994) analyzed the seed protein profiles of 19 accessions representing 7 of the 9 sections of the *Arachis* genus, and concluded that Section *Arachis* is phylogentically closest to Section *Erectoides* followed by *Procumbensae, Heteranthae* (formerly *Ambinervosae*), *Caulorrhizae, Triseminatae* (formerly *Triseminalae*) and *Extranervosae*, respectively (Singh et al., 1994).

First bootstrap consensus rooted phylogenic trees based on nuclear rDNA internal transcribed spacers (ITS) of *Arachis* species were constructed and published online in Genetic Resource and Crop Evolution on June 9, 2010 (Wang et al. 2011b). To exclude the possible interference of overlapping peaks in trace files, ITS PCR products were directly sequenced and only the unambiguous sequences were used for analysis. The results essentially support the current sub-generic classification of the *Arachis* genus (Krapovickas and Gregory, 1994), but some exceptions do exist. Notably, sections *Erectoides* and *Rhizomatosae* are not grouped together. According to the analysis, Sections *Extranervosae, Heteranthae* and *Triseminatae* are most primitive, whereas Section *Arachis* is most advanced, with Sections *Caulorrhizae, Erectoides, Procumbentes, Rhizomatosae,* and *Trierectoides* intermediate in evolutionary terms, in relation to the genus *Stylosanthes*, when it is used as the outgroup (Wang et al. 2011b).

Later, on Oct 6, 2010, phylogenetic trees for the *Arachis* genus were published using nuclear ITS and plastid *trn*T-*trn*F sequences (Friend et al., 2010). The study revealed that Section *Extranervosae* species are the first emerging lineage in the *Arachis* genus, followed by Sections *Triseminatae* and *Caulorrhizae,* and two terminal lineages: species from Sections *Erectoides, Heteranthae, Procumbentes, Rhizomatosae,* and *Trierectoides* form an intermediate clade, and Section *Arachis* species constitute a more advanced clade consisting of 2 clades (B-, D- genome and 2n=18 species and A-genome species form separate clades, and the A-genome clade is advanced). The difference in the placement of *Heteranthae* by the two research groups (Wang et al. 2011; Friend et al., 2010) is caused by use of different materials. In an unpublished study, when a different accession of *A. pusilla* was used, it appeared in the intermediate clade (C.T. Wang, unpublished data). This raises questions about the real identity of the two materials.

4. *ARACHIS* SPECIES AS ORNAMENTAL GROUNDCOVER

4.1. *A. glabrata*

Native to Brazil, Argentina and Paraguay, *A. glabrata* is also called rhizoma peanut in the United States of America. The species is highly persistent and rhizomatous.

4.1.2. Major Characteristics

A. glabrata adapts well to subtropical or warm climate. Once established, it shows good tolerance to drought, and grows and persists in a series of well-drained soil types ranging from sands to clays. Sand or organic matter should be added to heavy clay soils to improve drainage, if this soil type cannot be avoided. *A. glabrata* has had mixed success in high pH soils in Florida, USA. It grows best in full sun and partial shade. Low temperature is detrimental to rhizomes, and may even kill them. Prines (1983) recommended that rhizoma peanut should be grown in areas with no less than -10°C temperatures. Cold tolerance of different cultivars might, however, vary considerably. In cultivation, *A. glabrata* has grown successfully in places from near the equator to about 30ºN and 30ºS, with average annual temperature of ~ 20°C to 26°C (Cook et al., 2005). It can survive severe winters in north Florida and southern Georgia (Sotomayor-Rios and Pitman, 2001). *A. glabrata* is best suited to sites receiving 1000-2000 mm annual rainfall (Cook et al., 2005).

4.1.3. As Ornamental Groundcover

Rhizoma peanut, *A. glabrata*, has potential landscape uses as groundcover in home landscapes, road medians, driveways, parking lot islands, golf courses, along berms, septic tank mounds, and canal banks (R. E. Rouse, E. M. Miavitz and F. M. Roka, personal communication). The abundance of bright showy flowers produced each day is the superior ornamental attributes of Arblick and Ecorturf (Prine et al., 2010). Studies from around the world have shown that *A. glabrata* is more resistant to winter kill, spider mite damage, and nematode issues than pinto peanut (NRCS, 2008).

4.1.4. Cultivar Releases

Commercial cultivar releases of *A. glabrata* are listed in Table 4. But in earlier years, primary focus of evaluations was for forage use; hence, most of the cultivars were not bred for groundcover. Low growth rate would be an advantage if rhizoma perennial peanut was used in landscape settings as groundcover or turf (NRCS, 2008). For example, cultivars such as Arbrook and Florigraze grow tall and are mostly used for forage, rather than for ornamental. With lower growing characteristics, Arblick and Ecorturf are two cultivars released for ornamental or forage use. The only selections of peanut currently recommended as a perennial groundcover or turf In Florida, USA are Ecoturf, Arblick, Brooksville 67 and

Brooksville 68 (NRCS, 2008). As an ornamental groundcover, Ecoturf has already gained acceptance by the commercial landscape industry (Prine et al., 2010). This cultivar was least sensitive among six rhiozma peanut germplasm lines in above-ground growth to extended photoperiod in the field, which perhaps explains its superior performance as low maintenance turf selection that needs infrequent mowing (Prine et al., 2010).

Table 4. Commercial cultivar releases of *A. glabrata* and *A. pintoi*

Cultivars	Places (year) released	Source
Arb	Florida, USA (*ca.* 1960)	*A. glabrata* (PI 118457, CPI 58110)
Arblick	Florida, USA (2008)	*A. glabrata* (PI 658528)
Arbrook	Florida, USA (1986)	*A. glabrata* (PI 262817)
Brooksville 67 ("Waxy leaf")	Florida, USA (2002)	*A. glabrata* (GKP 9553, PI 262801)
Brooksville 68 ("pointed leaf")	Florida, USA (2002)	*A. glabrata* (NRCS #9056068)
Ecoturf	Florida, USA (2008)	*A. glabrata* (PI 658529)
Florigraze (GS-1)	Florida, USA (1978)	*A. glabrata* (PI 421707, a chance hybrid?)
Pointed leaf	Florida, USA	*A. glabrata* (CPI 93483, PI 231318)
Prine	Australia (Queensland) (1995)	*A. glabrata* (CPI 93483, PI 231318)
Reclaim	South Africa (1987)	*A. glabrata* (PI 118457, CPI 58110)
UF Tito	Florida, USA (2008)	*A. glabrata* (PI 262826)
UF Peace	Florida, USA (2008)	*A. glabrata* (PI 658214)
Latitude 34	Texas, USA (2009)	*A. glabrata* (PI 658497)
Alqueire-1	Rio Grande do Sul, Brazil (1998)	*A. pintoi* (BRA-037036)
Amarillo	Australia (1987)	*A. pintoi* (GK 12787, PI 337361, PI 338314, PI 338447, CPI 58113, CIAT 17434, IRI 2270, IRFL 4222, I 44457)
Amarillo MG-100	Brazil (1994)	*A. pintoi* (GK 12787)
Belmonte	Brazil (1999)	*A. pintoi* (BRA-031828)
Golden Glory	Hawaii (1996)	*A. pintoi* (origin unknown)

Table 4. (Continued)

Cultivars	Places (year) released	Source
Itacambira	Brazil	*A. pintoi* (W 34b, BRA-031143, CIAT 22160, ATF 2320, IRFL 7133)
Mani Forrajero	Panama (1997)	*A. pintoi* (GK 12787 + BRA-012122)
Mani Forrajero Perenne	Colombia (1992)	*A. pintoi* (GK 12787, CIAT 17434)
Mani Mejorador	Costa Rica (1994)	*A. pintoi* (GK 12787)
Pico Bonito	Honduras (1993)	*A. pintoi* (GK 12787)
Porvenir	Costa Rica (1998)	*A. pintoi* (BRA-012122, CIAT 18744, CPI 133550, ATF 495)
Arachis pintoi cv. Amarillo (阿玛瑞罗平托花生)	China (2003)	*A. pintoi* (L86-96)
A. pintoi cv Reyan No.12 (热研 12 号平托落花生)	China	*A. pintoi* (CIAT 22160)

Source: For *A. glabrata*, see http://www.tropicalforages.info/key/Forages/ Media/Html/Arachis_ glabrata.htm, and French et al. Establishment and management of ornamental perennial peanuts, SS-AGR-19(Available at: http://polkhort.ifas.ufl.edu/documents/publications/Establishment%20 and%20Mng%20of%20PP.pdf), Quesenberry et al. (2010) and Muir et al. (2010); for *A. pintoi*, see http://www.tropicalforages.info/key/Forages/Media/Html/ Arachis_pintoi.htm, Agricultural crops germplasm resource information system (http://tcgris.catas.cn/news_detail.aspx?id=576), Wu (2004) and Bai et al. (2006).

4.2. *A. pintoi*

As a section *Caulorhizae* member indigenous to central Brazil, *A. pintoi* was first collected by Professor G.C.P. Pinto in 1954. Now it is grown throughout the wet tropics and subtropics, from 0 to 1800 meters above sea level. Cultivated *A. pintoi* can be found in Australia, Argentina, The United States, Central America, Southeast Asia and South China (Kartika et al., 2009).

4.2.1. Major Characteristics

In central Brazil where it originated, *A. pintoi* has generally been seen growing on red, low to moderate fertility, sandy loam river-bottom soils of high Al saturation, especially in low areas, which are wet to flooded during the wet season (Cook and Crosthwaite, 1994). Wet season rainfall in the region is around 2000 mm, while a further 200 mm falling is received during the rest period of the year. Growing under trees up to 5 m in height in these places, *A. pintoi* can only get little sunlight, showing high level of shade tolerance (Cook and Crosthwaite, 1994).

When cultivated, *A. pintoi* is, in general, well adapted to soils of various textures; but poorly structured medium to heavy clays, viz., the so-called "puggy" soils, might be an exception (Cook and Crosthwaite, 1994). *A. pintoi* prefers moderate to high fertility but can survive in infertile soils. While *A. pintoi* only has low to moderate tolerance to salinity, it can tolerate high levels of Mn and Al (Cook and Crosthwaite, 1994). The species develops well on soils with pH values of ~4.3 (in water)-7.2 (in 1:5 water), though growth rate reduces on soils with lower than 5.4 pH values (Cook and Crosthwaite, 1994). Adapted well to moderately drained, seasonally inundated or poorly drained soils (Cook and Crosthwaite, 1994), *A. pintoi* shows tolerance to periods of waterlogging. As might be expected, cultivated *A. pintoi* also has good tolerance to heavy (70%-80%) shade. Over 650 mm annual rainfall is adequate, but higher than 1500 mm per year is best. *A. pintoi* can survive dry seasons of 3-4 months (Meng, 2012; Kartika, Susila and Reyes, 2009).

4.2.2. As Ornamental Groundcover

As a prostrate, stoloniferous, perennial herb with bright yellow flowers and green tetrafoliate leaves, *A. pintoi* can be used for multipurpose. In addition to forage uses, it is also a good cover crop for citrus, banana, oil palm, coffee, macadamia and "harts of palm" plantations in humid environments of tropics and subtropics (Argel, Kerridge and Pizarro, 1997). Though generally established slowly, once established, it rarely needs watering even under drought conditions, and due to its competitiveness resulting from high shade tolerance and dense mates of rooted stolons, the need for weed control is minimal, thereby reducing management costs (Argel, Kerridge and Pizarro, 1997). When cultivated with small shrubs and other herb flowers, it is better to dig a 30-50 cm wide isolation zone to protect the shallow rooted plants from being deprived of water and soil nutrients. Similarly, in orchards, *A. pintoi* should be planted outside the 70-100 cm diameter circles under the trees to avoid its adverse competitive effects on the trees (Meng, 2012).

In northern Queensland, Australia, *A. pintoi* is used for bank stabilization and as an ornamental groundcover in shaded or erosion-prone areas (Cook and Crosthwaite, 1994).

Studies conducted in Florida, USA by Abdul-Baki, Bryan and Klassen (2002) demonstrated that *A. pintoi* cv. Amarillo is a suitable ornamental plant along roadsides and highway ramps. The extensive root system and thick foliage cover of *A. pintoi* prevent soil erosion even on the steep slopes typically encountered with highway ramps (Abdul-Baki et al., 2002). Apart from the advantages mentioned above, *A. pintoi* has greatly reduced fertilizer needs, and can tolerate heavy traffic and wheel pressure by landscaping equipment, making it a sustainable groundcover with almost no maintenance requirements. This species routinely produces a prolific and attractive stand of yellow flowers set against an aesthetically pleasing green background that persist year-round (Abdul-Baki et al., 2002).

Amarillo was introduced into China in 1986 by Forage Grass Research Center, Chinese Academy of Tropical Agricultural Sciences (Liu and Luo, 1999). In Yunan province, it was found to grow very well in Simao, and it showed good persistence in Xiaosao (Zhou et al., 2001). Researchers from Fujian Academy of Agricultural Sciences introduced *A. pintoi* into China in 1990. The cultivar passed China National Forage Cultivars Pre-release Evaluation Test in December 2003, and renamed as *Arachis pintoi* cv. Amarillo (阿玛瑞罗平托花生). It is widely planted in eco-orchards as cover crop and used for landscaping programs in Fujian province, China (Wu, 2004). In South China, CIAT 22160, an introduction from Centro

Internacional de Agricultura Tropical (CIAT), Columbia, performed consistently well in pot culture experiments, greenhouse evaluation and field tests, when compared with 4 other CIAT *A. pintoi* introductions (CIAT 18750, CIAT 17434, CIAT 18748 and CIAT 18744) and cv. Amarillo. In a 5-year-period (1999-2004) multi-location production trial in Hainan, Guangdong, Guangxi and Yunnan, it was superior to the check cultivar Amarillo in productivity. It was released as *A. pintoi* cv. Reyan No.12 (热研 12 号平托落花生) by Tropical Crops Genetic Resources Institute, Chinese Academy of Tropical Agricultural Sciences. *A. pintoi* cv. Reyan No.12 begins to flower in less than a month after planting, and flowering will last for more than 10 months. It has been and can be widely used as pasture plant, cover crop in plantations, groundcover along roadsides and on steep slopes, and ornamental plant in South China (Bai et al., 2006).

4.2.3. Cultivar Releases

Commercial cultivars of *A. pintoi* that have been released are summarized in Table 4. Amarillo might be the most famous *A. pintoi* cultivar in the world. Introduced into Australia as CPI 58113 in 1972, *A. pintoi* PI 338314 was released commercially in 1987, registered as cv. Amarillo in 1990 (Cook and Crosthwaite, 1994), and proved to be a commercial success (Sotomayor-Ríos and Pitman, 2001). It was exported to Colombia, Indonesia, Malaysia, Thailand, Sri Lanka, several Pacific countries and China (Cook and Crosthwaite, 1994; Liu and Luo, 1999; Wu, 2004; Sotomayor-Rios and Pitman, 2001).

4.3. Other Species

4.3.1 A. duranensis

In China, *A. duranensis* was included in the "List of recommended commonly used ornamental plant species in Hainan province" by the Hainan provincial government (2011), and is being used as groundcover in Guangxi, Fujian and Taiwan provinces (Li et al., 2009; Zheng et al., 2007; Chen, 2006). In Chaozhou city, Guangdong province, Zhang and Zheng (2008) compared the turf performance of *A. pintoi* cv. Amarillo and *A. duranensis*, and found that *A. duranensis* was superior to Amarillo in coverage, leaf-flower ratio, grow rate, leaf area and chlorophyll contents. Now, *A. duranensis* is widely used as ornamental groundcover in several parks in Shenzhen city, Guangdong. The species is one of five herb plant species widely used in urban landscape in Guangzhou city, Guangdong (Wang et al., 2007). In Yunan province, the species shows good resistance to poisonous gases, and grows well in Kunming Steel and Iron Company Forest Garden Nursery (102°10'-102°37'E, 24°31'-25°6'N), provided that suitable measures are taken to protect it from frost injury (for example, using non-woven fabrics) (Li and Li, 2010). *A. duranensis* also performs well in Xishuangbanna Tropical Garden (Chen, 2006). In a green roof study conducted by Architectural Services Department, Hong Kong (2009), *A. duranensis* grew very well, and based on interim observations it was listed as one of the three recommended groundcover species. It also works well if grown in a hanging pot (Tan and Pan, 2004).

The species showed good saline tolerance, and can be used in south China seaside regions (Lin, 2004). Under polyethylene glycol (PEG) simulated drought stress, *A. duranensis* was more tolerant than *A. pintoi* cv. Amarillo, *A. glabrata* (an accession from Guangxi) or *A.*

pintoi cv. Reyan No.12 (Lv, 2006). Luo et al. (2009) reported that as compared with *A. pintoi* cv. Amarillo, *A. duranensis* had equal productivity, smaller leaves, bigger flowers and better cold hardiness. The species proved to be tolerant of acid soil, drought and shade stress, and can be used for ecological restoration of eroded soil in tea fields in Fujian (Luo et al., 2009).

4.3.2. A. repens

As an ornamental, *A. repens* is still underutilized in places outside its native environment. During a collection trip in Brazil, Dr C.E. Simpson noted that *A. repens* (locally known as 'grama amendoim' or groundnut grass) was growing in domestic and commercial gardens. At the collection site, *A. repens* PI 338277 was mainly used for lawn planting (Cook and Crosthwaite, 1994).

CONCLUSION

Up to now, most of the studies related to wild peanut species have been conducted by peanut breeders, rather than workers on flowers and turf science, with aims to enhance the productivity and stress resistances of *A. hypogaea* more often and in some cases to transfer desirable quality traits from wild peanut species (Wang et al., 2013b). Several *A. hypogaea* cultivars of interspecific origin, including four peanut cultivars derived from an intersectional cross, have been released (Wang et al., 2013b). Great efforts have also been devoted to the development of forage use peanut cultivars with wide adaptation, high stress resistance, good persistence and high forage quality and yields. Using wild *Arachis* species as ornamental groundcover cultivars is a new breeding objective. As in forage peanut, ease of planting is also important for ornamental peanut. Slow growth rate, high and multiple stress resistance, low requirements for fertilization are necessary to minimize maintenance costs.

It should be noted that only a small portion of wild peanut is used in ornamental groundcover research. There are multiple numbering and naming systems for wild peanut. Intraspecific variation also exists. Again, the subgeneric classification system for the genus *Arachis* has several versions (see Wang et al., 2013b), and the names of some wild peanut species even along with their sectional affiliations have changed. Similar modifications may happen in the future. Therefore, the collector's abbreviation and number should always be provided to clarify the identity of the peanut materials.

Some wild peanut species only produce a limited number of seeds. In extreme cases, no seeds are set. For the exchange of peanut materials, seeds are always more convenient than vegetative tissues. Methods to enhance seed yield should be developed. Ironically, wild peanut species that produce no seeds are advantageous, as the breeding process can be shortened. No seed, no segregation. In addition, heterosis is readily accessible in such cases. Consequently, besides introduction and pure line selection, induced mutagenesis and intrasectional hybridization are logical immediate choices for breeding methods. Intersectional hybridization may also find some uses in the future. Updated information on taxonomy, crossability and phylogeny will undoubtedly facilitate varietal improvement of ornamental peanut.

ACKNOWLEDGEMENTS

The study was financially supported by the earmarked fund for China Agricultural Research System (CARS-14). Special thanks are due to Dr Massimiliano Dematteis, Instituto de Botánica del Nordeste, for kindly permitting us to quote "the Key for identifying individual sections in the *Arachis* genus" originally published in Spanish in Bonplandia, 8(1-4):20 in 1994, and in English in Bonplandia 16 (supl.): 33-34 in 2007.

REFERENCES

Abdul-Baki, A.A., Bryan, H.H., Klassen, W. and Codallo, M. (2002). Propagation and establishment of perennial peanuts for ground covers along roadsides and highway ramps. *Proceedings of Flarida State Horticultural Society.* 115: 267-272.

Ahmed, E.M. and Fletcher, D.L. (1977). Response of peanut protein spun fibers to applied stresses. Peanut Science. 4(1): 22-26.

Architectural Services Department, Hong Kong. (2009). Green roof & vertical greening applications in government projects. URL:http://www.archsd.gov.hk/media/ 11617/c2154.pdf (accessed May 20, 2013).

Argel, P.J., Kerridge, P.C. and Pizarro, E.A. (1997). *Arachis pintoi*: a multipurpose legume for sustainable land use. URL: http://www.internationalgrasslands.org/files/igc/ publications/1997/2-19-083.pdf.

Bai, C.J., Liu, G.D., He, H.X., Zhou, J.S. and Wang, D.J. (2006). Breeding and Selection of *Arachis pintoi* cv. Reyan 12. *Chinese Journal of Tropical Crops.* 27(2):45-49.

Chen, W.J. (2006). *Arachis duranensis*: a high quality groundcover. *Tropical Agricultural Science and Technology.* 29(1):38-40.

Cook, B.G. and Crosthwaite, I.C. (1994). Utilization of *Arachis* species as forage. In J. Smartt (Ed.), The groundnut crop: a scientific basis for improvement. pp.624-665. London: Chapman & Hall.

Cook, B.G., Pengelly, B.C., Brown, S.D., Donnelly, J.L., Eagles, D.A., Franco, M.A., Hanson, J., Mullen, B.F., Partridge, I.J., Peters, M. and Schultze-Kraft, R. (2005). Tropical Forages: an interactive selection tool. [CD-ROM], CSIRO, DPI&F(Qld), CIAT and ILRI, Brisbane, Australia. URL: http://www.tropicalforages.info/ (accessed May 20, 2013).

Fletcher, D.L. and Ahmed, E.M. (1977). Spinning of peanut protein fibers. *Peanut Science.* 4(1):17-21.

Friend, S.A., Quandt, D., Tallury, S.P., Stalker, H.T. and Hilu, K.W. (2010). Species, genomes, and section relationships in the genus *Arachis* (Fabaceae): a molecular phylogeny. *Plant Systematics and Evolution.* 290:185-199.

Grabiele, M., Chalup, L., Robledo, G. and Seijo, G. (2012). Genetic and geographic origin of domesticated peanut as evidenced by 5S rDNA and chloroplast DNA sequences. *Plant Systematics and Evolution.* 298:1151-1165.

Guan, A.H., Zhang, J.F and Zhang, C.J. (2006). New type renewable protein fibers. *Synthetic Fibers.* (6):24-27.

Hainan Provincial Government. (2011). List of Recommended commonly used ornamental plant species in Hainan province. URL: http://www.hainan.gov.cn/data/zfwj/2011/02/3488/ (accessed May 20, 2013).

Husted, L. (1933). Cytological studies of the peanut *Arachis*. I. Chromosome number and morphology. *Cytologia*. 5:109-117.

Husted, L. (1936). Cytological studies of the peanut *Arachis*. II. Chromosome number, morphology and behaviour and their application to the origin of cultivated form. *Cytologia*. 7:396-423.

Kartika, J.G., Susila, A.D. and Reyes, M.R. (2009). Review of literature on perennial peanut (*Arachis pintoi*) as potential cover crop in the tropics. Kumpulan Makalah Seminar Ilmiah Perhortl. pp. 391-399. URL: http://repository.ipb.ac.id/bitstream/handle/123456789/33099/PRO2009_JGK.pdf?sequence=5 (accessed May 24, 2013).

Kerridge, P.C. and Hardy, B. (Eds.) (1994). Biology and agronomy of forage *Arachis*. Cali, Columbia: Centro Internacional de Agricultura Tropical (CIAT). 209pp.

Krapovickas, A. and Gregory, W.C. (1994). Taxonomy of the genus *Arachis* (Leguminosae). *Bonplandia*. 8:1-186.

Krapovickas, A. and Gregory, W.C. (2007). Taxonomy of the genus *Arachis* (Leguminosae). Translated by Williams D.E. and Simpson, C.E. Bonplandia. 16(supl.):1-205.

Lavia, G.I., Fernandez, A. and Seijo, J.G. (2008). Cytogenetic and molecular evidences on the evolutionary relationships among *Arachis* species. In: A.K. Sharma, A. Sharma (Eds.), Plant genome: biodiversity and evolution. Vol. 1E. Pahnerogams – Agiosperm. Enfield, NH, USA: Science Publishers. pp. 101-134.

Li, M.Z. and Li, J. (2010). Cultivation and propagation techniques for *Arachis duranensis*. *China Horticulture Digest*. (2):95-96.

Li, X.J., Qiu, S., Zhao, J., Zhang, G.P., Liu, S.H. and Li G.Z. (2009). Investigation and application of ornamental plants in Guangxi. *Guihaia*. 29(5):635-639.

Lin, G.S. (2004). Major anti-wind and alkali-resisting landscape plants of south China's seaside region. *Forest Inventory and Planning*. 29(3):78-81.

Liu, G.D. and Luo, L.J. (1999). Forage plant germplasm resources for tropical regions in China. Beijing: China Agricultural University.

Luo, X.H., Zhong, Z.M., Zhan, J., Chen, Q.S. and Huang, Y.B. (2009). Forages useful in ecological restoration of eroded soil in tea fields in Fujian. *Subtropical Soil and Water Conservation*. 21(4):45-48.

Lv, L.X. (2006). Study on drought resistance of 4 forages of *Arachis* spp. Master's Thesis. South China University of Tropical Agriculture. Danzhou, Hainan, China.

Meng, G.P. (2012). Cultivation and management of gold peanut – *Arachis pintoi*. *China Flowers and Horticulture*. (4):30-31.

Moretzsohn, M.C., Gouvea, E.G., Inglis, P.W., Leal-Bertioli, S.C., Valls, J.F.and Bertioli, D.J. (2013). A study of the relationships of cultivated peanut (*Arachis hypogaea*) and its most closely related wild species using intron sequences and microsatellite markers. *Annals of Botany*. 111(1):113-126.

Muir, J.P., Butler T.J., Ocumpaugh, W.R. and Simpson, C.E. (2010). 'Latitude 34', a perennial peanut for cool, dry climates. *Journal of Plant Registrations*. 4(2):106-108.

National Resources Conservation Service (NRCS). (2008). Plant materials fact sheet. Rhizoma perennial peanut (*Arachis glabrata*) – The perennial peanut for urban conservation in Florida. URL: http://taylor.ifas.ufl.edu/ documents/FlaPerennialPeanut FactSheet_08.pdf (accessed June 6, 2013).

Nelson, A.D., Samuel, M., Tucker, J., Jackson, C. and Stahlecker-Roberson, A. (2006). Assessment of genetic diversity and sectional boundaries in tetraploid peanut (*Arachis*). *Peanut Science*. 33:64-67.

Nóbile, P.M., Gimenes, M.A., Valls, J.F.M. and Lopes, C.R. (2004). Genetic variation within and among species of genus *Arachis*, section *Rhizomatosae*. *Genetic Resources and Crop Evolution*. 51(3):299-307.

Prine, G.M. (1983).'Rhizoma peanut': perennial warm season forage legume. In: Proceedings of the XIV Internacional Grassland Congress, Lexington, Kentucky. Westview Press.

Prine, G.M., French, E.C., Blount, A.R., Williams, M.J., Quesenberry, K.H. (2010). Registration of Arblick and Ecoturf rhizoma peanut germplasms for ornamental or forage use. *Journal of plant registrations*. 4(2):145-148.

Quesenberry, K.H., Blount, A.R., Mislevy, P., French, E.C., Williams, M.J. and Prine, G.M. (2010). Registration of 'UF Tito' and 'UF Peace' rhizoma peanut cultivars with high dry matter yields, persistence, and disease tolerance. *Journal of plant registrations*. 4(1):17-21.

Robledo, G. and Seijo, G. (2010). Species relationships among the wild B genome of *Arachis* species (section *Arachis*) based on FISH mapping of rDNA loci and heterochromatin detection: a new proposal for genome arrangement. *Theoretical and Applied Genetics*. 121(6):1033-1046.

Rouse, R.E., Miavitz, E.M. and Roka, F.M. Guide to using rhizomal perennial peanut in the urban landscape. URL: http://edis.ifas.ufl.edu/ep135 (accessed May 23, 2013).

Simpson, C.E. (1995). Evolution and variability in the *Arachis* genus. In D.G. Cummins (Ed.), Proceedings of peanut CRSP workshop. pp. 307-318. Washington D.C.

Singh, A.K. (1985). Cytogenetic analysis of wild species of *Arachis*. Project Report (1978-82). ICRISAT. Patancheru. India. p. 71.

Singh, A.K., Gurtu, S. and Jambunathan, R. (1994). Phylogenetic relationships in the genus *Arachis* based on seed protein profiles. *Euphytica*. 74:219-225.

Singh, A.K., Simpson, C.E. (1994). Chapter 4. Biosystematics and genetic resources. In: J. Smartt (Ed.) The groundnut crop: a scientific basis for improvement. pp. 96-137. London: Chapman & Hall.

Smartt, J. (1990). Grain legumes: evolution and genetic resources. Cambridge University Press. 392pp.

Smartt, J., Gregory, W.E. and Gregory, M.P. (1978). The genomes of *Arachis hypogaea* 1. Cytogenetic studies of putative genome donors. *Euphytica*. 27:665-675.

Sotomayor-Ríos, A. and Pitman, W. (Eds.). (2001). Tropical forage plants: development and use. CRC Press.

Stalker, H.T. (1991). A new species in section *Arachis* of peanut with a D genome. *American Journal of Botany*. 78:630-637.

Stalker, H.T. and Simpson, C.E. (1995). Germplasm resource in *Arachis*. In H.E. Pattee and H.T. Stalker (Eds.), Advances in peanut science. pp. 14–53. Stillwater: American Peanut Research and Education Society.

Tan, G.Y. and Pan, Z.Q. (2004). Cultivation and uses of perennial groundnut, *Arachis duranensis*. *Guangxi Tropical Agriculture*. (1):46-47.

USDA, ARS, National Genetic Resources Program. Germplasm Resources Information Network - (GRIN) [Online Database]. National Germplasm Resources Laboratory, Beltsville, Maryland. URL: http://www.ars-grin.gov/cgi-bin/npgs/html/tax_search.pl (accessed 19 May 2013).

Wan, S.B. (Ed.). (2003). Peanut cultivation science in China. Shanghai: Shanghai Science and Technology Press. 647pp.

Wang, C.T., Tang, Y.Y., Wang, X.Z., Wu, Q. (2011a). Breeding peanut for forage and cover crop uses. pp. 506-513. In: S.L. Yu, C.T.Wang, Q.L. Yang, D.X. Zhang, X.Y. Zhang, Y.L. Cao, X.Q. Liang and B.S. Liao (Eds.), Peanut genetics and breeding science in China. Shanghai: Shanghai Science and Technology Press.

Wang, C.T., Wang, X.Z., Tang, Y.Y., Chen, D.X., Cui, F.G., Zhang, J.C. and Yu, S.L. (2011b). Phylogeny of *Arachis* based on internal transcribed spacer sequences. Genetic Resources and Crop Evolution. 58(2):311-319.

Wang, C.T., Tang, Y.Y., Wang, X.Z., Wu, Q., Gao, H.Y., Feng, T., Su, J.W., Yu, S.T., Fang, X.L., Ni, W.L., Jiang, Y.S., Qian, L. and Hu, D.Q. (2012). Mutagenesis: a useful tool for the genetic improvement of the cultivated peanut (*Arachis hypogaea* L.). In: R. Mishra (Ed.), Mutagenesis. pp. 1-12. Crotia: Intech.

Wang, C.T., Zhang, J.C., Tang, Y.Y., Guan, S.Y., Wang, X.Z., Wu, Q., Shan, L., Zhu, L.G., Su, J.W. and Yu, S.T. (Eds.). (2013a). Genetic improvement of peanut. Shanghai: Shanghai Science and Technology Press.

Wang, C.T., Tang, Y.Y., Wang, X.Z., Wu, Q., Yang, Z., Song, G.S., Gao, H.Y., Yu, S.T., Ni, W.L., Yang, T.R., Li, M. and Qian L. (2013b). Utilizing wild incompatible *Arachis* species in the genetic improvement of the cultivated peanut (*A. hypogaea* L.). Chapter 11. In: J.N. Govil (Ed.), Recent developments in biotechnology. Vol. 2. Plant biotechnology. Studium Press LLC. (in press)

Wang, Z., Ouyang, C.J., Luo, Y.Y. and Wang, R.J. (2007). Application of indigenous plants to urban landscape in Guangzhou. *Subtropical Plant Science*. 36(4):33-37.

Woodroof, J.G. (1983). Nonfood uses for peanuts. In J.G. Woodroof (Ed.), Peanut: production, processing and products. 3rd ed. pp. 361-367. Westport, Connecticut: AVI Publishing Company, Inc.

Wu, H.N. (2004). Pintoi peanut passed China National Forage Cultivars Pre-release Evaluation Test. *Fujian Agricultural Science and Technology*. (1):42.

Xing, S.Y. (Ed.). (2005). Handbook of products in chemical industry: textile fibers. Beijing: Chemical Industry Press. 512pp.

Yu, S.L., Wang, C.T., Yang, Q.L., Zhang, D.X., Zhang, X.Y., Cao, Y.L., Liang, X.Q. and Liao, B.S. (Eds.). (2011). Peanut genetics and breeding science in China. Shanghai: Shanghai Science and Technology Press. 565pp.

Zhang, Z.X. and Zheng, Y.Z. (2008). Study on the turf quality of two species of *Arachis*. *Hubei Agricultural Sciences*. 47(9):1053-1055.

Zheng, X.L., Ying, C.Y., Xu, G.Z. and Huang, Y.B. (2007). Cultivation techniques and uses of *Arachis duranensis*. *Fujian Agricultural Science and Technology*. (6):32-33.

Zhou, Z.W., Zhong, S., Kui, J.X., He, Z.S. and Kuang, C.Y. (2001). High quality forage legume - Amarillo peanut. *Sichuan Grassland*. (3):43-47.

Reviewed by Prof. Yue Huang, Qingdao Academy of Agricultural Sciences

In: Flowers
Editors: Teodor Berntsen and Kaj Alsvik

ISBN: 978-1-62808-798-7
© 2013 Nova Science Publishers, Inc.

Chapter 3

UNDERSTANDING THE ROLE OF PIGMENTS IN FLOWERS

María Gabriela Lagorio[*]

INQUIMAE/Dpto. de Química Inorgánica, Analítica y Química Física. Facultad de
Ciencias Exactas y Naturales. Universidad de Buenos Aires. Ciudad Universitaria.
Pabellón II, 1er piso, Buenos Aires, Argentina

ABSTRACT

Betalains, anthocyanins, carotenoids and chlorophylls, are the main pigments present
in flowers. Anthocyanins and betalains are mutually exclusive and they are never found
together in a plant. The main function of the pigments in flowers is giving them colour
and thus to provide an attraction signal to pollinators. The hue of flowers, however, not
only depends on the nature of their pigments but it is also influenced by the presence of
co-pigments, the cellular pH and the incorporation of metal ions. Other factors like light
conditions and temperature also affects the spectral distribution of the light reflected by
the pigments and as consequence the flower color. Several auhors have also observed
fluorescence emission from betalains in flowers. The reflected light and the light emitted
as fluorescence by flowers pigments may be analysed in terms of their relevance in
biosignalling towards pollinators. For this survey, not only the spectral distribution of the
light given off from the flowers, but also the sunlight spectrum and the sensitive response
of pollinators photoreceptors should be considered. Some flowers pigments show also a
defensive function acting as toxics to predator insects. It was found that carotenoids are
related to flowers flavor as they act as precursors of chemicals which are responsible for
the fragrance of some flowers.

1. CHEMICAL NATURE OF FLOWERS PIGMENTS

Four major types of pigments are present in flower petals: betalains, flavonoids,
carotenoids and chlorophylls [Grotewold, 2006]. Each pigment is formed as the result of a

---

[*] Fax: 54 11 4576 3341, Phone: 54 11 4576 3378 int.106, e-mail: mgl@qi.fcen.uba.ar.

complex series of biochemical reactions and the synthesis of each of them is independent of the production of the others. Pigments are responsible for the attractive beauty of flowers but they also play different relevant roles in the interaction of plants with other species in the environment.

1.1. Betalains

Betalains are water soluble aromatic indole derivatives (Figure 1). They are synthesised from the aminoacid tyrosine [Grotewold, 2006]. Their name comes from the Latin denomination of the beet (*Beta vulgaris*) from which they were first extracted [Kujala et al., 2001].

In the group of betalains we can find betacyanins, usually violet-red and betaxanthins, normally yellow to orange [Waterman, 1988; Salisbury and Ross, 1991; Strack et al., 1993; Strack et al., 2003].

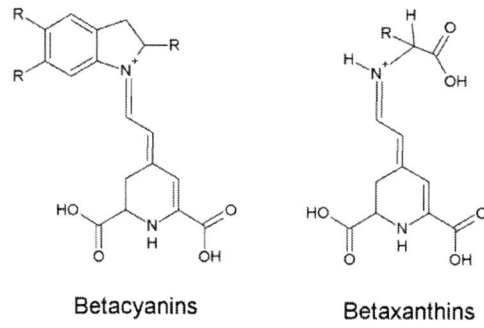

Betacyanins Betaxanthins

Figure 1. Chemical structure of Betalains.

Betalains are present in plants of the family *Caryophyllales*. Examples of flowers containing red or violet betacyanins are *Portulaca grandiflora* (portulaca), *Mirabilis jalapa* (beauty of the night), *Amaranthus* (amaranths), *Bougainvillea* and *Dianthus caryophyllus* (carnations). Yellow betaxanthins are present in flowers of *Portulaca grandiflora and Lampranthus* among others [Heuer et al., 1994]

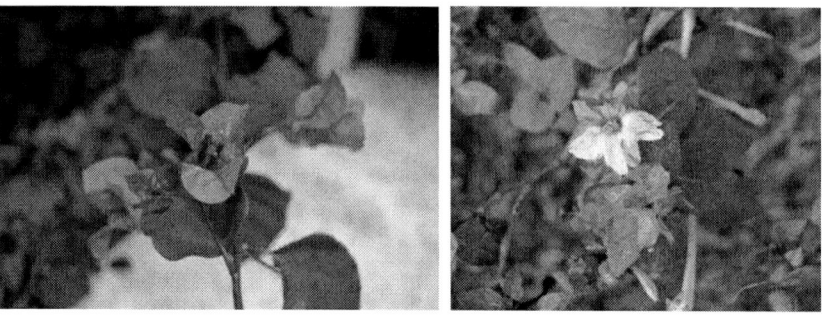

Figure 2. *Bouganvillea* and *Mirabilis jalapa*. Flowers containing betalains.

Betalains are stored in flowers vacuoles as glycosides. In plants where betalains are present, anthocyanins are absent and this mutual exclusion seems to be general. Betalains are known to have anti-oxidant and anti-inflammatory effects in humans [Allegra et al., 2005; Escribano et al., 1998] and they are stable over the pH range from 3 to 7, allowing its use as colorant in low acidity foods [Jackman and Smith, 1996]. An excelent review on betalains properties, uses and sources may be found in [Azeredo, 2009].

1.2. Flavonoids

Flavonoids (from the Latin word *flavus*: yellow, golden) are water soluble compounds, stored in flowers vacuoles. Despite the name etymology, their colors vary from pale-yellow to blue. Chalcones, flavones, flavonols and anthocyanins belong to this group. While chalcones, flavones and flavonols are yellow pigments (Figure 3), anthocyanins (Figure 4) may be orange to blue, changing their colors with pH.

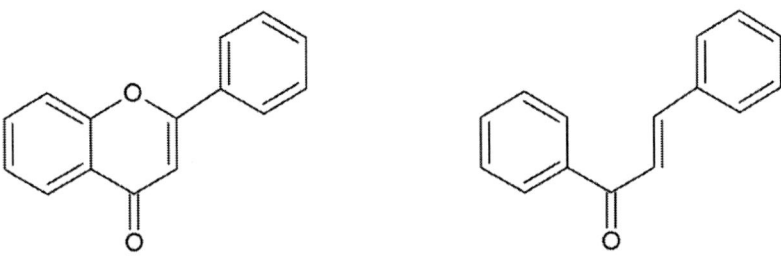

Basic structure of Flavones   Basic structure of Chalcones

Basic structure of Flavonols

Figure 3. Basic structure for flavones, chalcones and flavonols.

The six major known anthocyanins are pelargonidin, cyanidin, peonidin, delphinidin, petunidin and malvidin. The larger the number of hydroxyl groups in their molecule, bluer the color they present. [Tanaka et al., 2008]. In particular, the name Anthocyanin comes from the Greek *anthos*: flower and *kianos*: blue [Castañeda-Ovando, 2009]. Anthocyanins are basically anthocyanidines bonded to a sugar moity which confers the high water solubility. Anthocyanines are synthesised in flowers as part of the flavonoid pathway [Grotewold, 2006] and they are easily degraded once isolated from the plant [Giusti and Wrolstad, 2001; Giusti and Wrolstad, 2003].

Differing from betalains, anthocyanins do not contain nitrogen (Figure 4). They are widely distributed in flowers and provide them with different colors: orange, red, purple, blue.

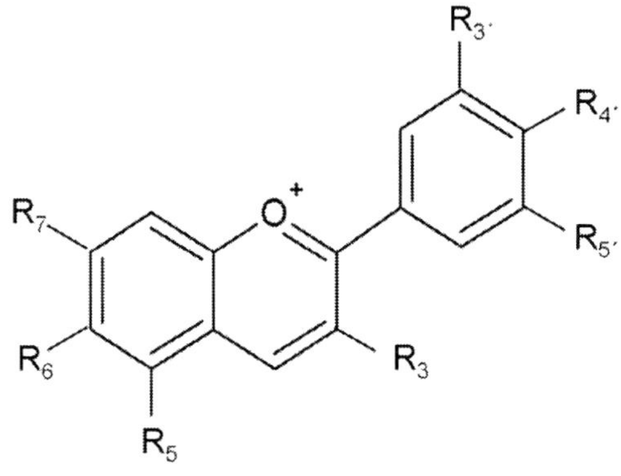

Figure 4. Anthocyanins general structure. Where R_5, R_6, R_7, R_3, $R_{3'}$, $R_{4'}$, and R_5 are H or OH or OCH_3. *Pelargonidin*, R3′: -H, $R_{4'}$: -OH, $R_{5'}$: -H, R_3: -OH, R_5: -OH, R_6: -H, R_7: -OH; *Cyanidin*, R3′: -OH, $R_{4'}$: -OH, $R_{5'}$: -H, R_3: -OH, R_5: -OH, R_6: -H, R_7: -OH; *Peonidin*, R3′: -OCH$_3$, $R_{4'}$: -OH, $R_{5'}$: -H, R_3: -OH, R_5: -OH, R_6: -H, R_7: -OH; *Malvidin*, R3′: -OCH$_3$, $R_{4'}$: -OH, $R_{5'}$: -OCH$_3$, R_3: -OH, R_5: -OH, R_6: -H, R_7: -OH; *Petunidin*, R3′: -OH, $R_{4'}$: -OH, $R_{5'}$: -OCH$_3$, R_3: -OH, R_5: -OH, R_6: -H, R_7: -OH; *Delphinidin*, R3′: -OH, $R_{4'}$: -OH, $R_{5'}$: -OH, R_3: -OH, R_5: -OH, R_6: -H, R_7: -OH.

Anthocyanins color depends on pH and for this reason they can be used as pH indicators. In acidic media they are red. By increasing the pH, they turn pink becoming purple around pH= 7. In alkaline solution, they become green-yellow. Anthocyanins have ability to form complexes with metal ions, which also influences the observed colors. [Boulton, 2001]. They are present in flowers of *Petunia* [Spelt et al., 2006] *Pelargonium* [Mitchell et al., 1998], *Hibiscus* [Aishah et al. 2013], *Hydrangea macrophylla* [Kondo et al., 2005], and *Rhododendron indicum* [Iriel and Lagorio, 2009] among others.

Figure 5. *Pelargonium* and *Hibiscus*. Flowers containing anthocyanins.

Anthocyanins are harmless compounds and they are used as natural water soluble colorants [Pazmiño-Durán et al., 2001].

1.3. Carotenoids

Carotenoids are lipid soluble pigments. The name derives from the Latin word *carota* (carrot), a root vegetable very rich in the pigment beta carotene. They possess a polyisoprenoid structure as a common chemical feature (Figure 6).

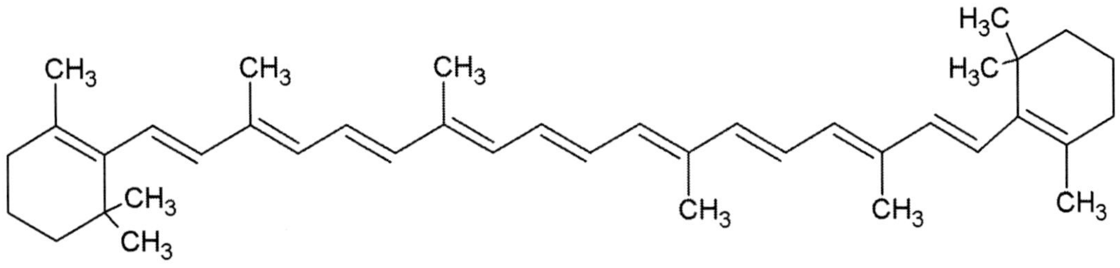

Figure 6. Beta carotene, an example of carotenoids structure.

The carotenoids family is very large consisting of more than 600 compounds. Carotenoids are present in flowers in small packets called chromoplasts. They are antioxidants and they are precursors for the biosynthesis of vitamin A in animals [Fraser and Bramley, 2004]. Carotenoids give yellow to orange colour to flowers being present in marigolds (*Tagetes*), daffodil (*Narcissus*), *Freesia*, *Gerbera*, *Rosa*, *Lilium* and *Calendula* [Grotewold, 2006]. Carotenoids can coexist either with anthocyanins or with betalains.

Figure 7. *Tagetes* and *Rosa*. Flowers containing carotenoids.

A complete review on carotenoids structure, properties and functions can be found in [Britton, 1995].

1.4. Chlorophylls

Apart from anthocyanins, carotenoids and betalains, chlorophylls may be also present in flowers. Chlorophylls are lipid soluble pigments (Figure 8) contained in chloroplasts and they are the main pigment involved in photosynthesis. Green leaves are the natural place where the photosynthetic process occurs, however, several authors have also reported the presence of photosynthetically active chloroplasts in mature flowers [Dueker and Arditti, 1968; Vu et. al. 1985; Weiss and Halevy, 1989]. Active chloroplasts, similar in size and structure to those from green leaves have been particularly reported for Pink Petunia *hybrida* (cv Hit Parade

Rosa) [Weiss et al., 1988] and for carnations corollas [Vainstein and Sharon, 1993] In these cases, chlorophyll content of corollas increased during the flower development, reaching a maximum just before anthesis and it was suggested that photosynthetic activity in corolla chloroplasts could be involved in the corolla development.

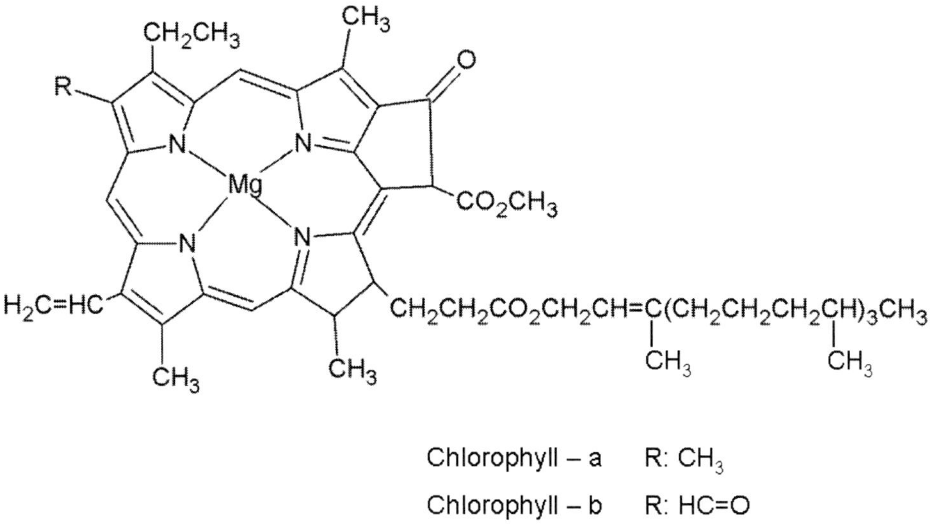

Chlorophyll – a   R: CH$_3$
Chlorophyll – b   R: HC=O

Figure 8. Chlorophyll structure.

Figure 9. Petunia. Flower containing chlorophyll.

2. THE ROLE OF PIGMENTS IN VISUAL SIGNALLING

Pigments play a relevant role in the ecology of plants as they are responsible for flower colors producing optical signals to attract pollinators and organisms that participate in seed dispersion [Weiss, 1991 and 1995]. To understand the optical phenomena of vision it is important to notice that "Color is not actually a property of light or of objects that reflect light. It is a sensation that arises within the brain" as stated by Timothy H. Goldsmith, [Goldsmith, 2006]. The perceptions that different animals have of the same object are different, depending on the ability of their eyes photoreceptors to be excited by light of different wavelengths. So, to construct a color image in brain, a combination of several events should take place: light emission from an illuminating source impinging on an object, reflection of part of the incident light by the illuminated object, excitation of the eyes photoreceptors of the observer by the light reflected by the object, conversion of the photoreceptors excitation to a signal, and transduction of the optical signal in the brain to create an image. The first two events depend exclusively on the spectral distribution of the light source and the reflectance spectra of the object, but the last two depend on each observer.

2.1. Light Reflection of Flowers and its Evaluation as a Biosignal

To understand how the light reflected from flowers pigments acts on pollinator vision, a physico-mathematical approach may be used. In this context, the relative amount of light absorbed by the eyes photoreceptors per unit time is a very meaningful quantity. These quantities are known in literature as Quantum catches and to estimate them, the reflectance spectrum of the observed object, $R(\lambda)$, the photon flow of the illumination beam, $I(\lambda)$, and the photoreceptor spectral sensitivity $S_i(\lambda)$ should be known as a function of wavelength (λ) [Iriel and Lagorio, 2010]. As the reflectance represents the fraction of the excitation light reflected by the flower, it is possible to calculate the total amount of light reflected by it as the product $R(\lambda)I(\lambda)$. To estimate then the amount of photons absorbed by the pigments of a given photoreceptor i, the product $R(\lambda)I(\lambda)$ should be multiplied by $S_i(\lambda)$. To count the total amount of photons absorbed by each photoreceptor, integration of the last product from 300 to 700 nm is performed (Pi) [Iriel and Lagorio, 2010 a and b].

$$P_i = \int_{300}^{700} R(\lambda)S_i(\lambda)I(\lambda)\,d\lambda \tag{1}$$

Different types of photoreceptors are sensitive to different regions of electromagnetic spectrum.

The relative amount of light absorbed by the eyes photoreceptors per unit time (Quantum catches) are then calculated directly as Pi divided by the total amount of photons absorbed by each photoreceptor, coming from the excitation source [Kelber et al., 2003; Vorobyev et al., 2001]:

$$Q_i = \frac{P_i}{\int_{300}^{700} S_i(\lambda)I_i(\lambda)\,d\lambda} \tag{2}$$

Quantum catches are calculated in this way as "relative" quantities, so that upon a change in the intensity of the illumination beam, Q_i is not expected to vary significantly. This kind of constancy represents the adaptation of each photoreceptor to the photon flux [Smithson, 2005]. In fact, photoreceptors decrease their sensitivity in direct proportion to the illumination received [Hurley, 2002] (Weber's Law).

Regarding the signals emitted to the environment, plants face a real dilemma: they should display an attraction to pollinators but they should remain hidden and camouflaged (if possible) against hamful organisms as parasites.

Pollinators analyse the signals from flowers, mainly the visual and optical messages. They process these inputs in their brains to elaborate a subjective perception on the external world [Chitka and Raine, 2006]. As stated before, a crucial point to understand how visual cues act is to realise that the subjective perception created by light signals in the brain, depend on the observer. In fact, a flower observed by a bee, by a bird and by a human will produce different color sensations in each of the brains.

Bees have 3 types of color receptors with peaks at 340, 430 and 540 nm [Chittka, 1996; Briscoe and Chittka, 2001], Humans are also trichromats but with photoreceptors maximum sensitivities at 440 nm (B), 543 nm (G) and 566 nm (R) [Smith and Pokorny, 2003]. Birds photoreceptors array is the most evolved among the natural species [Bowmaker 2008]. They have four spectrally different cone pigments displaying absorption maxima at 543-571 nm, 497-510 nm, 430-463 nm and at very short wavelength (violet or ultra violet sensitive), 362-426 nm [Hart, 2001; Maier, 1994] (Figure 10).

As an example, quantum catches calculated for a white flower observed by a human and observed by a bee are shown in Figure 11.

The three photoreceptors in humans are almost equally excited when looking at the considered flower giving us the sensation of "white" color. In the case of bees, in contrast, the UV receptor is poorly excited compared with the blue and green ones. As we do not know how the transduction process transforms the photosignal into an image in the brain, we cannot infer, from the calculated quantum catches, what color the bees actually see. However, we can assure this will be chromatic because their three receptors are not equally excited.

Not only absolute colors but also contrasts between flowers and background are relevant for pollinator attraction. The contrast to the background (C_i) may be calculated dividing Pi by the total amount of photons absorbed by receptor i from the background (P_{ib}) (equation 3 and 4) [Kelber et al., 2003; Chittka and Kevan, 2005]:

$$C_i = \frac{P_i}{P_{ib}} \tag{3}$$

$$P_{ib} = \int_{300}^{700} S_i(\lambda)\, I_i(\lambda)\, R_b(\lambda)\, d\lambda \tag{4}$$

where R_b is the background reflectance.

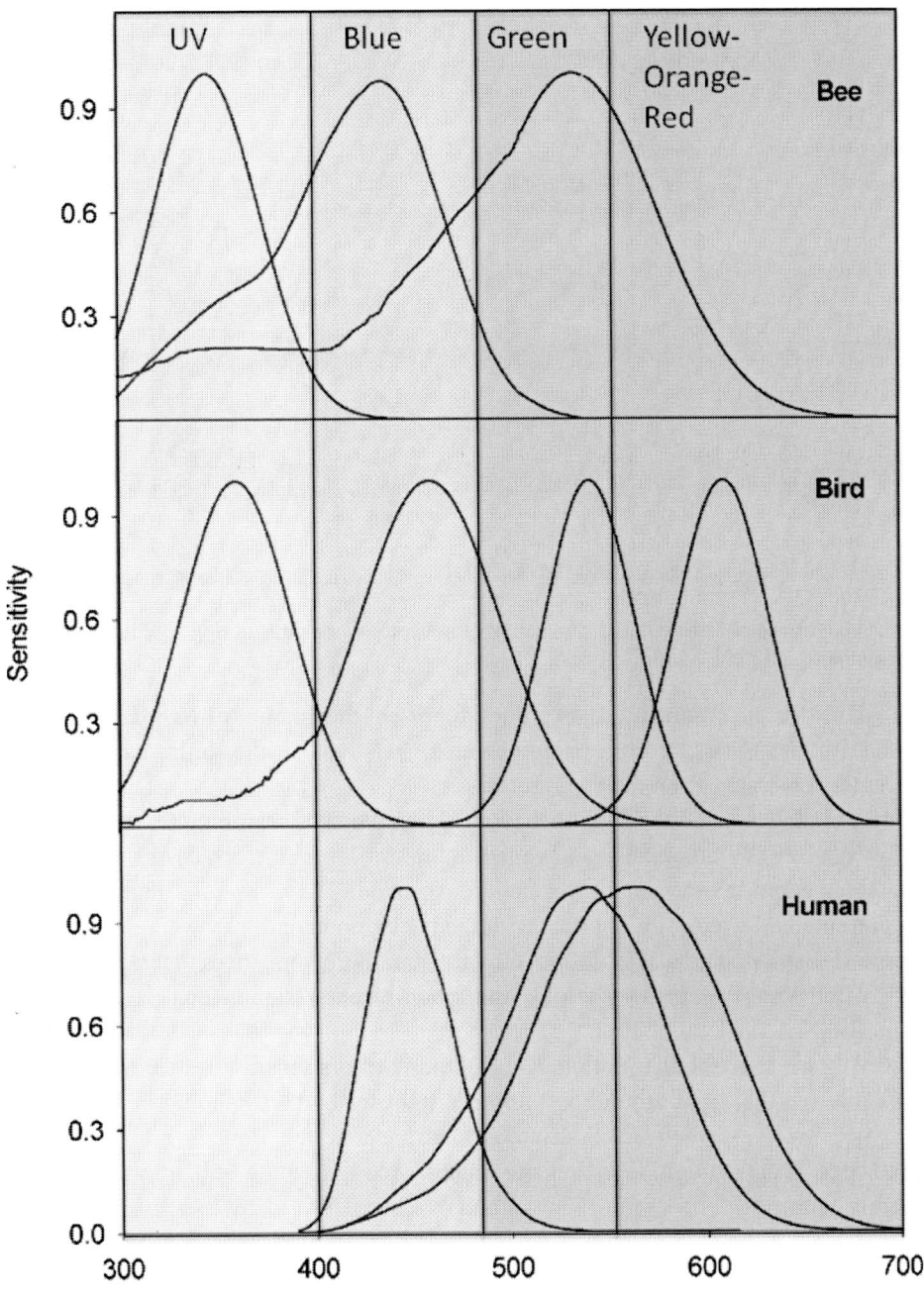

Figure reproduced with permission from Iriel and Lagorio, 2010b.

Figure 10. Photoreceptor sensitivities for the honeybee *Apis melifera*, for the bird *Leiothrix lutea* and for humans.

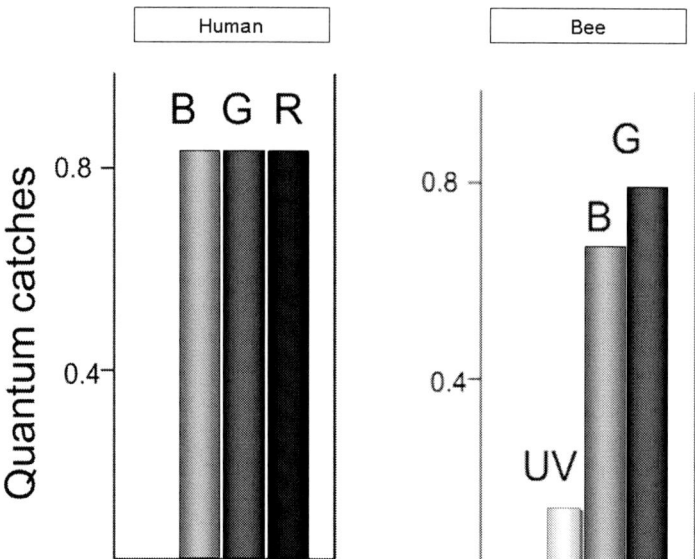

Figure 11. Quantum catches calculated for each photoreceptor when considering a white flower. Left figure for humans and right figure for bees.

The values for the contrast to background represent the amount of light coming from the object that is absorbed by the photoreceptors, relative to the light captured by the photoreceptors emerging from the background.

Spaethe et al. evaluated the flight behavior of bumble bees searching for artificial flowers with different sizes and colors. They have found that for large flowers, search time correlated with the color contrast between flower and background. They observed that white and red flowers presented a lower color contrast with the background and they were detected by bees with more difficulty. They also found that for small flowers, bees decrease their flight velocity to find forage and they use only the green receptor signal for detection instead of color contrast [Spaethe et al., 2001].

Observing the spectral distribution of the photoreceptors sensitivities for bees (Figure 10) it may be inferred that light around 650 nm (red light for humans) should be poorly or non-detected by bees. Nevertheless, some authors assure that red flowers are not completely invisile to bees. If the sunlight is sufficiently intense, red light up to 650 nm will stimulate the bee receptor with peak at 540 nm [Chittka and Wasser, 1997]. However, bees require more time to find red flowers than blue ones [Spaethe et al., 2001].

2.2. Fluorescence in Flowers

Some pigments contained in the flowers are fluorescent and they are responsible for the luminescence detected in petals. Betaxanthins absorbs light in the blue and display emission in the green region. Gandía Herrero et al. have reported visible fluorescence from betaxanthins contained in petals of *Mirabilis jalapa*, *Portulaca grandiflora* and *Lampranthus productus* [Gandía-Herrero et al., 2005a and 2005b]. Ono et al. have also reported the fluorescent emission from aurones (a kind of flavonoids) in *Antirrhinum majus* [Ono et al.,

2008]. Thorp has reported fluorescence emission from the flowers nectar [Thorp et al., 1975] and Iriel and Lagorio have detected fluorescence from anthocyanins in *Rhododendron indicum* petals [Iriel and Lagorio, 2009; Iriel and Lagorio, 2010b].

2.3. Reflectance versus Fluorescence

The fact of displaying fluorescence emission is not enough for considering this signal as a call for pollinators in flowers. In fact, to assess the relevance of fluorescence as a biosignal towards pollinators, a quantitative comparison between the number of fluorescent photons and the number of reflected photons should be done, taking into account, additionally, the optical response of the photoreceptors in pollinators eyes. Such a study was performed by Iriel and Lagorio [Iriel and Lagorio, 2010a]. Different sorts of flowers have been analysed in that work: yellow flowers containing betaxanthins (*Lampranthus productus*, *Portulaca grandiflora*, and *Bougainvillea spectabilis*), yellow flowers containing aurones (*Antirrhinum majus* and *Limonium sinuatum*), pink, purple, and blue petals containing anthocyanins (*Bellis perennis and Eustoma grandiflorum*) and white flowers (*Petunia nyctaginiflora*, *Bougainvillea spectabilis*, *Antirrhinum majus, O. thyrsoides, Citrus aurantium, Bellis perennis, Limonium sinuatum,* and *Eustoma grandiflorum*). Flowers' ovaries (*O. thyrsoides*) and stigmas (*Citrus aurantium*) were also studied. As tempting as it may be as an attractive speculation on the signal fluorescence might offer the pollinators in locating flowers, the experiments have shown that fluorescence signals were negligible compared to reflected light in all the cases. The results pointed to refuse a visual function of flower fluorescence. Peter Kevan had already questioned in 1976 the ecological relevance of floral fluorescence [Kevan, 1976], a prediction that was supported experimentally in 2010 [Iriel and Lagorio, 2010a].

2.4. Influence of Co-Pigments, Metal Ions and pH on Flowers Color

Packaging of pigments with other copigments and metal ions are also found in flowers petals.

Strikingly, the anthocyanin that makes roses red was found in blue cornflowers. In 1915, Willstätter et al. assigned this color variation to a different pH in the flowers petals [Willstätter et al., 1915]. In 1919, Shibata et al. suggested instead, an association of the pigment with metal ions [Shibata et al., 1919] and in 1931, Robinson et al. reported a combination with another pigment [Robinson and Robinson, 1931]. Finally, in 2005, Shiono et al. gave experimental evidence (X-ray crystal structure of the cornflower pigment) supporting that the brilliant blue color was due to "a complex of six molecules each of anthocyanine and flavones, with one ferric ion, one magnesium and two calcium ions" [Shiono et al., 2005].

Blue colors in flowers are mainly due to anthocyanines and most of these blue colors can be ascribed to metal-complex anthocyanin. Momonoi et al., studied petals of *Tulipa gesneriana* cv. Murasakizuisho and they have found that differences in iron content were responsible for variations in color from blue to purple [Momonoi et al., 2009].

The acidity of the petals cells also has an important influence on color, particularly when anthocyanins are present. In general, the cells of blue flowers are more alkaline than red ones.

This fact is consistent with the red color of anthocyanins in acid media and the blue color in basic media respectively.

When studying *Hydrangeas*, however, it is found that despite containing anthocyanins, they produce pink flowers when the soil pH is about 6 and produce blue flowers when the soil acidity is increased (pH around 5.5), opposing to the general trend exposed before. The explanation for this behavior is the increased availability of aluminum at acidic pH. In fact, absorbed aluminum forms complexes with anthocyanins changing the color from pink to blue. In this way, the availability of aluminum nullifies the effect of acidity.

Some flowers cells change their pH with aging. For instance, young Morning glory flowers, which contain anthocyanins as the main pigment, have a pH around 6.5 and display a pink color. Mature flowers of the same species shift their pH to about 7.5 and their color turn blue. When the flowers are about to close, their color becomes pink again due to a decrease in the petal pH to around 6.

There are also flowers containing anthocyanins that do not change their colors neither upon changing the soil pH, nor during aging. *Rhododendron indicum* flowers behave this way. The reason for this fact is that *Rhododendron* flowers are buffered and their pH is not changed by modifications in the soil pH or during aging. The pH of *Rhododendron* flowers is under genetic control with a little influence of the environment [Griesbach, 1987].

Figure 12. *Rhododendron* indicum flowers.

2.5. Influence of Light Conditions and Temperature

Light and temperature also affect flower color. At high light intensities, the production of sugar is increased in plants due to a rapid photosynthesis rate. Sugar binds to anthocyanins molecules stabilizing their color. In addition, as a protective reaction towards excess light, plants increase the synthesis of anthocyanins. Both these factors (the increase in the amount of sugar and in the amount of anthocyanins) led to brighter petals colors. Low temperatures also have beneficial effects enhancing colors. Cold weather slows down the plant growth and the amount of sugar used in respiration. So, under low temperatures, more sugar is available to be bound to anthocyanins. In this way, cold weather combined with high light promotes intense colors in petals containing anthocyanins.

3. THE ROLE OF PIGMENTS IN DEFENSE

In addition to their visual function, some pigments may play a role in defense against predators. Gronquist et al., for instance, have found that the dearomatized isoprenylated phloroglucinols, (DIPs), which are present in high concentration in the anthers and ovarian wall of the flower *Hypericum calycinum,* are toxic and repellent to a caterpillar *Utetheisa ornatrix.* DIPS are responsible for an UV pattern which is probably detected by the insects. In this way, some floral UV pigments might have both visual and defensive functions. DIPs have been also proposed as protective substances of female reproductive structures in other plants, as hops (*Humulus lupulus*) [Gronquist et al., 2001].

It was also proposed that Anthocyanins in petals (which have well known visual functions) may also act as a defense against excess light, pathogen agents or insects attacking plants. In an interesting experiment on petunia flowers displaying white and colored sectors, Johnson and Berhow found that corn earworm larvae ate significantly less colored sectors than white sectors of the flowers. They have also found that anthocyanines mixtures extracted from petunia flowers, added to the insect diets, reduced corn earworm larvae weights [Johnson and Berhow, 2008].

4. THE ROLE OF PIGMENTS IN FLAVOR SIGNALS

Pigments are usually associated with the color in flowers; less is known about their relationship with the smell of flowers.

In some species, carotenoids give degraded products which are responsible for the flower fragrance. These products are usually potent scents. In plants, carotenoids are constantly synthesised and degraded by enzymatic oxidative cleavage giving as result a set of terpenoid products collectively known as apocarotenoids [Beisel et al. 2010]. The degradation products include abscisic acid and strigolactones, and other volatile products, which are used for aromas, flavors and fragrances in industry. Some apocarotenoids, as β-ionone, also play a role in plant-insect interactions [Beisel et al. 2010].

Several flowers, appreciated for their fragrance contain degradation products from carotenoids in their petals. From *Boronia*, for example, twenty degradation compounds from carotenoids have been detected and have been assigned to be responsible for its perfume. *Boronia* is a plant from *Rutaceae* family, whose flowers have an intense aroma. The plant is original from Australia and it is widely used in the flavor industry [Lawrence, 1999; Weyerstahl et al., 1994; Weyerstahl et al., 1995; Davies and Menary, 1983].

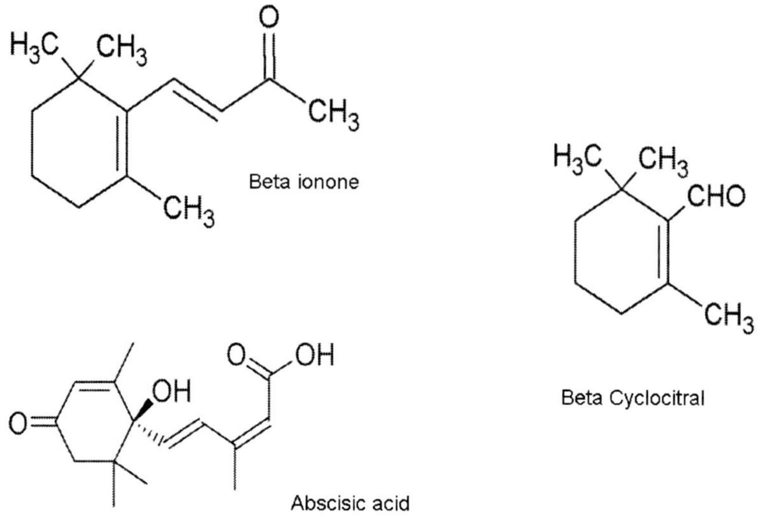

Figure 13. Examples of degradation compounds from carotenoids (apocarotenoids).

Osmanthus fragrans is a flower native to China with a fruity-floral apricot aroma. Its essential oil is very expensive and used in selected perfumes and flavors. Carotenoid derived compounds are responsible for the greater part of this scent [Gogiya et al., 1986; Ishiguro et al., 1957].

Rose aroma has been appraised for its delicious perfume from Romans time. It is well known that citronellol, which is not a carotenoid derivative, is the major constituent of Rose essential oil. However, carotenoids derivatives, in spite of being minority, contribute greatly. In fact, in 1960-1970's the minor components of the rose fragrance were reported and afterwards the contribution of each component to aroma was analysed, based on the "Odor unit concept" of Guadagni [Guadagni et al., 1966]. This quantity was defined as the ratio between the concentration of the compound and the odor threshold. So, as the value of the odor unit becomes higher, the contribution of the substance to the aroma increases. Ohloff reviewed the aroma contribution of the major and minor components in the Bulgarian rose oil and he found that the minor constituents beta-damascenone and beta-ionone, both degradation products from carotenoids, provided a significant majority of the odor contribution [Ohloff, 1994].

The flower of *Crocus sativa* has a red stigma that is prized as spice (saffron) and as colorant [Basker and Negbi, 1983; Cadwallader and Back, 1997]. The saffron color is due to degraded carotenoids (as crocin and crocetin) [Pfander and Rychener, 1982], and its flavor comes from oxidation products of carotenoids (mainly safranal and picrocrocin). Pfander and Schurtenberger have proposed that both the compounds responsible for the odour and the compounds responsible for the aroma are the result of the bio-oxidative cleavage of Zeaxanthin (Figure 14) [Pfander and Schurtenberger, 1982; Pfander and Witter 1975a and 1975b].

The main volatile component of Saffron is [2,6,6-Trimethylcyclohexa-1,3-dien-1-carboxaldehyde] (usually called Safranal- Figure 15), which is formed by de-glucosylation of picrocrocin [Alonso et al., 1996; Tarantilis and Polissiou, 1997].

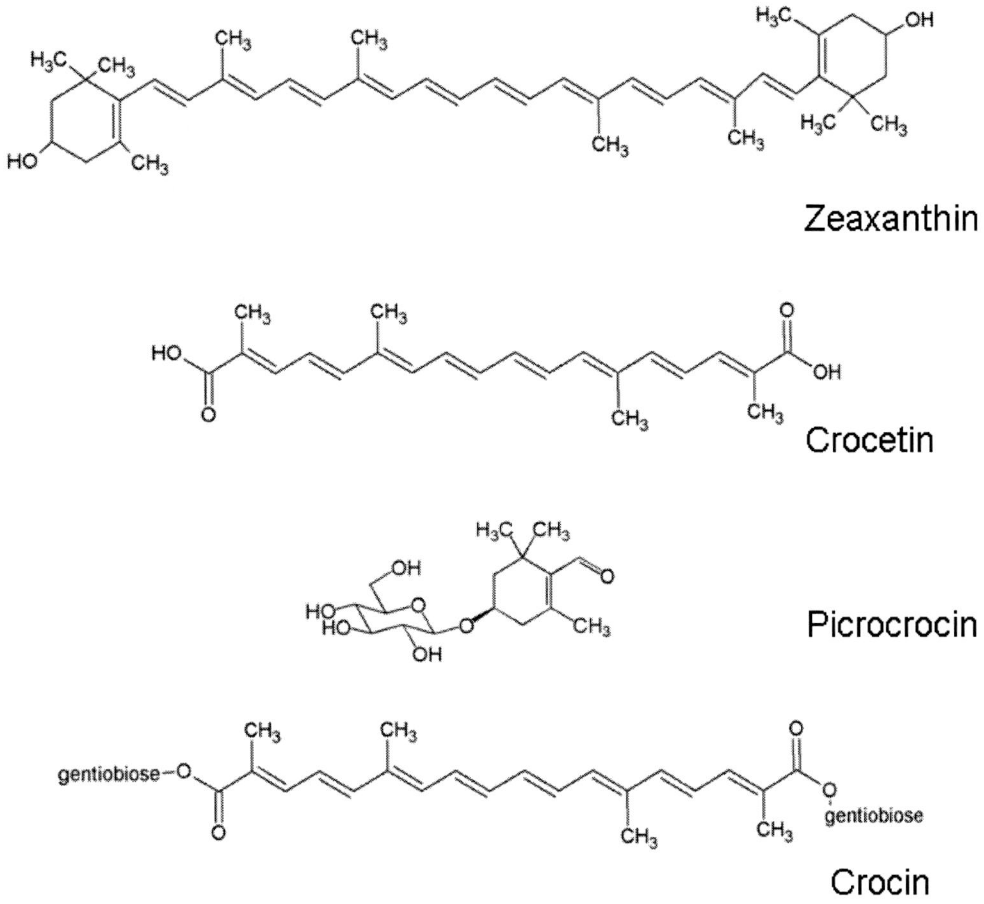

Zeaxanthin

Crocetin

Picrocrocin

Crocin

Figure 14. Zeaxanthin and degradation products.

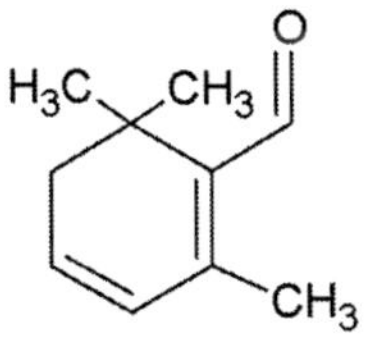

Safranal.

Figure 15. Chemical structure for Safranal, the major component in Saffron volatiles.

The products of carotenoids degradation are key odour compounds not only in many flowers but also in aged tobacco, in black tea and in many fruits including grapes.

ACKNOWLEDGEMENTS

The authors are grateful to the University of Buenos Aires (Project UBACyT 20020100100814) for the financial support.

REFERENCES

Aishah, B., Nursabrina, M., Noriham, A., Norizzah, A. R., Mohamad Shahrimi, H. Anthocyanins from Hibiscus sabdariffa, Melastoma malabathricum and Ipomoea batatas and its color properties. *Int. Food Res. J.*, 2013, 20, 827-834.

Alonso, G. L., Salinas, M. R., Esteban-Infantes, F. J., Sanchez Fernandes, M. A. Determination of safranal from saffron (Crocus sativus; L.) by thermal desorption-gas chromatography. *J. Agric. Food Chem.*, 1996, 44, 185.

Allegra, M., Furtmüller, P. G., Jantschko, W., Zederbauer, M., Tesoriere, L., Livrea, M. A., Obinger, C. Mechanism of interaction of betanin and indicaxanthin with human myeloperoxidase and hypochlorous acid. *Biochem. Bioph. Res. Co.*, 2005, 332, 837-844.

Azeredo, H. M. C. Betalains: properties, sources, applications, and stability - a review. *Int. J. Food Sci. Tech.* 2009, 44, 2365-2376.

Basker, D.; Negbi, M. Uses of Saffron. *Econ. Bot.*, 1983, 37, 228-236.

Beisel, K. G., Jahnke, S., Hofmann, D., Köppchen, S., Schurr, U., et al. Continuous turnover of carotenes and chlorophyll a in mature leaves of Arabidopsis revealed by $^{14}CO_2$ pulse-chase labeling. *Plant Physiol.*, 2010, 152, 2188-2199.

Boulton, R. The copigmentation of anthocyanins and its role in the color of redwine: A critical review. *Am. J. Enol. and Viticult.*, 2001, 52, 67-87.

Bowmaker J. K. Evolution of vertebrate visual pigments. *Vis Res*, 2008, 48, 2022-2041.

Briscoe, A. D. and Chittka, L. The evolution of color vision in insects. *Annu. Rev. Entomol.*, 2001, 46, 471-510.

Britton, G. Structure and properties of carotenoids in relation to function. *FASEB J.*, 1995, 9, 1551-1558.

Cadwallader, K. R., Back, H. H., Cai, M.; ACS Symposium Series, No. 166 (Spices), American Chemical Society, Pub., 1997, Washington DC, pp.66-79.

Castañeda-Ovando, A., Pacheco-Hernández, M. L., Páez-Hernández, M. E., Rodríguez, J. A., Galán-Vidal, C. A. Chemical studies of anthocyanins: A review. *Food Chem.*, 2009, 113, 859-871.

Chittka, L. Does bee color vision predate the evolution of flower color? *Naturwissenschaften*, 1996, 83, 136-138.

Chittka, L. and Wasser, N. Why red flowers are not invisible to bees. *Isr. J. Plant. Sci.*, 1997, 45,169-183.

Chittka, L. and Kevan, P. G. in Practical Pollination Biology; Dafni, A.; Kevan, P. G. and Husband B. C. (Eds), Enviroquest Ltd. 2005, Cambridge, ch. 4, pp. 157-196.

Chittka, L. and Raine, N. E. Recognition of flowers by pollinators, *Curr. Opin. Plant Biol.*, 2006, 9: 428-435.

Davies, N.W. and Menary, R. C. Volatile constituents of Boronia megastigma flowers, *Perfumer & Flavorist*, 1983, 9, 3-8.

Dueker, J., and Arditti, J. Photosynthetic CO, fixation by green Cymbidium (Orchidaceae) flowers. *Plant. Physiol*, 1968, 43, 30-132.

Escribano, J., Pedreño, M. A., García-Carmona, F., Muñoz, R. "Characterization of the antiradical activity of betalains from Beta vulgaris L. roots". *Phytochem. Anal.*, 1998, 9, 124-7.

Fraser P. D., Bramley, P. M. The biosynthesis and nutritional uses of carotenoids. *Prog. Lipid Res.*, 2004, 43, 228-265.

Gandía-Herrero, F, Garcia-Carmona, F, Escribano, J. Floral fluorescence effect. *Nature*, 2005a, 437, 334.

Gandía-Herrero, F, Escribano, J, García-Carmona, F. Betaxanthins as pigments responsible for visible fluorescence in flowers. *Planta*, 2005b, 222, 586-593.

Giusti, M. M., and Wrolstad, R. E. Unit F1.2.1–13. Anthocyanins. Characterization and measurement of anthocyanins by UV-Visible spectroscopy. In R. E. Wrolstad (Ed.), Current protocols in food analytical chemistry, John Wiley & Sons, 2001, New York, USA.

Giusti, M. M., Wrolstad, R.E. Acylated anthocyanins from edible sources and their applications in food systems. *Biochem. Eng. J.*, 2003, 14, 217-225.

Gogiya, V. T.; Kharebava, L. G., Gogiya R. V. and Gvatua, E. B. Composition of the volatile compounds in flowers of Osmanthus fragrans (Thunb.). *Lour. Rastit. Resur.*, 1986, 22, 243-248.

Goldsmith, T. H. What birds see. *Sci. Am.*, 2006, 69-75.

Griesbach, R. J. RHODODENDRON FLOWER COLOR: GENETIC/CULTURAL INTERACTION. Journal of American Rhododendron Society, 1987, 41. URL: http://scholar.lib.vt.edu/ejournals/JARS/v41n1/v41n1-griesbach.htm

Gronquist, M., Bezzerides, A., Attygalle, A., Meinwald, J., Eisner, M. and Eisner, .T. Attractive and defensive functions of the ultraviolet pigments of a flower (Hypericum calycinum). *PNAS*, 2001, 98, 13745–13750.

Grotewold, E. The Genetics and Biochemistry of Floral Pigments, *Annu. Rev. Plant. Biol.*, 2006, 57, 761-80.

Guadagni, D. G., Buttery; R. G., Harris, J. *J. Sci. Food. Agric.*, 1966, 17, 142-144.

Hart, N. S. The visual ecology of avian photoreceptors. *Prog Retin Eye Res*, 2001, 20, 675-703.

Heuer, S., Richter, S., Metzger, J. W., Wray, V., Nimtz, M., Strack, D. Betacyanins from bracts of Bougainvillea glabra. *Phytochemistry*, 1994, 37, 761-767.

Hurley, J. B. Shedding Light on Adaptation a relationship. *J. Gen. Physiol.*, 2002, 119, 125-128.

Iriel, A and Lagorio M. G. Biospectroscopy of *Rhododendron indicum* flowers. Non-destructive assessment of anthocyanins in petals using a reflectance-based method. *Photochem. Photobiol. Sci.*, 2009, 3, 337-344.

Iriel, A and Lagorio M. G. Is the flower fluorescence relevant in biocommunication? *Naturwissenschaften*, 2010 a, 97, 915-924.

Iriel, A. and Lagorio, M. G. Implications of reflectance and fluorescence of *Rhododendron indicum* flowers in biosignaling. *Photochem. Photobiol. Sci.*, 2010 b, 9, 342-348.

Ishiguro, T.; Koga, N., Nara, K. Components of the Flowers of Osmanthus fragrans II. Odorous Component Osmane and Acids. *Yakugaku Zasshi*, 1957, 77, 566-67.

Jackman, R. L. and Smith, J. L. Anthocyanins and betalains. In: Natural food colourants. G.F. Hendry & J.D. Houghton, Eds. Blackie Academic & Professional., 1996, London, UK, pp. 244-309.

Johnson, E T, Berhow, M. A. Colored and white sectors from star-patterned petunia flowers display differential resistance to corn earworm and cabbage looper larvae. *J Chem Ecol.* 2008, 34, 757-65.

Kelber, A, Vorobyev, M., Osorio, D. Animal Colour vision-behavioural tests and physiological concepts. *Biol. Rev.*, 2003, 78, 81-118.

Kevan, P. G. Fluorescent nectar. *Science*, 1976, 194, 341-342.

Kondo, T., Toyama-Katob, Y., Yoshida, K. Essential structure of co-pigment for blue sepal-color development of hydrangea. *Tetrahedron Lett.*, 2005, 46, 6645-6649.

Kujala, T., Loponen, J., Pihlaja, K. Betalains and phenolics in red beetroot (Beta vulgaris) peel extracts: extraction and characterization. *Z. Naturforsch.*, 2001, 56c, 343-348.

Lawrence, B., Progress in Essential Oils, Boronia, *Perfumer & Flavorist*, 1999, 24, 53-55.

Maier, E. J., Ultraviolet vision in a passeriform bird: from receptor spectral sensitivity to overall sensitivity in Leiothrix lutea. *Vision Res.*, 1994, 34, 1415-1418.

Mitchell, K. A., Markham, K. R., Boase, M. R. Pigment chemistry and colour of Pelargonium flowers. *Phytochemistry*, 1998, 47, 355-361.

Momonoi, K., Yoshida, K., Mano, S., Takahashi; H., Nakamori,. C., Shoji, K., Nitta, A., Nishimura, M. A vacuolar iron transporter in tulip, TgVit1, is responsible for blue coloration in petal cells through iron accumulation *The Plant Journal*, 2009, 59, 437-447.

Ohloff, G.; Scent and Fragrances, The fascination of odors and their chemical perspectives; translated by W. Pickenhagen and B. Lawrence, Springer-Verlag, Pub., 1994, Berlin – Heidelberg, pp.154-158.

Ono E, Fucuchi-Mizutani M, Nakamura N, Fukui Y, Yonekura-Sakakibara K, Yamaguchi M, Nakayama Tanaka Y, Sasaki N, Ohmiya A. Biosynthesis of plant pigments; anthocyanins, betalains and carotenoids. *Plant J.*, 2008, 54, 733-749.

Pazmiño-Durán, A. E., Giusti, M. M., Wrolstad, R. E., Glória, B. A. Anthocyanins from oxalis triangularis as potential food colorants. *Food Chem.*, 2001, 75, 211-216.

Pfander, H.; Schurtenberger, H. Biosynthesis of C_{20}-carotenoids in *Crocus sativus*. *Phytochemistry*, 1982, 21, 1039-1042.

Pfander, H.; Rychener, M. Separation of crocetin glycosyl esters by high-perfbrmance liquid chromatography. *J. Chromatogr.*, 1982, 234, 443-447.

Pfander, H.; Wittwer, F. Carotenoid glycosides. Part 2. Studies on the carotenoid composition of saffron. Helv. *Chim. Acta*, 1975a, 58, 1608-1620.

Pfander, H.; Wittwer, F. Carotenoid glycosides. Part 3. Studies on the carotenoid composition of saffron. Helv. *Chim. Acta* 1975b, 58, 2233-2236.

Robinson, G. M. and Robinson, R. *Biochem. J.,* 1931, 25, 1687-1705.

Salisbury, F. B. and Ross, C. W. *Plant Physiology* (4th ed.). Wadsworth Publishing, 1991, Belmont, California: pp. 325-326.

Shibata, K. Shibata, H. and Kashiwagi, I. *J. Am. Chem. Soc.*, 1919, 41, 208-220.

Shiono, M., Matsugaki, N., Takeda, K. Structure of the blue cornflower pigment, *Nature*, 2005, 791.

Smith, V. C. and Pokorny, J. Color matching and color discrimination. In: Shevel l SK, Ed. Science of color, 2nd edn. Elsevier, 2003, Oxford, pp 117-120.

Smithson, H. E. Sensory, computational and cognitive components of human colour constancy. *Philos. Trans. R. Soc. London, Ser. B*, 2005, 360, 1329-1346.

Spaethe, J., Tautz, J., Chittka, L. Visual constraints in foraging bumblebees: Flower size and color affect search time and flight behavior. *PNAS*, 2001, 98, 3898-3903.

Spelt, C, Quattrocchio, F., Mol, J., Koes, R. ANTHOCYANIN1 of Petunia Controls Pigment Synthesis, Vacuolar pH, and Seed Coat Development by Genetically Distinct Mechanisms. *The Plant Cell*, 2002, 14, 2121-2135.

Strack, D., Steglich, W., Wray, V. Betalains. In: Methods in Plant Biochemistry, v.8 Dey, P.M., Harborne, J.B. and Waterman, P.G. Eds., Academic Press, 1993, Orlando, pp. 421-450.

Strack, D., Vogt, T., Schliemann, W. Recent advances in betalain research. *Phytochemistry*, 2003, 62, 247-269.

Thorp R. W., Briggs D. L., Estes J. R., Erickson E. H. Nectar fluorescence under ultraviolet irradiation. *Science*, 1975, 189, 476-478.

Tanaka, Y., Sasaki, N., Ohmiya, A. Biosynthesis of plant pigments: anthocyanins, betalains and carotenoids. *The Plant Journal*, 2008, 54, 733-749.

Tarantilis, P. A.; Polissiou, M. Isolation and identification of the aroma constituents of saffron (*Crocus sativa*). *J. Agric. Food Chem.* 1997, 45, 459-462.

Vainstein, A. and Sharon, R. Biogenesis of petunia and carnation corolla chloroplasts: changes in the abundance of nuclear and plastid-encoded photosynthesis-specific gene products during flower development. *Physiol. Plantarum*, 1993, 89, 192-98.

Vorobyev, M., Marshall, J., Osorio, D., Hempel de Ibarra N., Menzel R. Colorful objects through animal eyes. *Color Res. Appl.*, 2001, 26, S214–S217.

Vu, J. C. V., Yelenosky, G., Bausher, M. G. Photosynthetic activity in the flower buds of Valencia orange (Citrus sinensis [L] Osbeck). *Plant Physiol.*, 1985, 78, 420-423.

Waterman, P. G. Alkaloid chemosystematics, G. A. Cordell, Ed. The Alkaloids-Chemistry and Pharmacology, Academic Press, 1988, San Diego, pp. 537–565.

Weiss, D., Schönfeld, M., Halevy, A. H. Photosynthetic Activities in the Petunia Corolla. *Plant Physiol.*, 1988, 87, 666-670.

Weiss, D. and Halevy, A. H. The role of light reactions in the regulation of anthocyanin synthesis in Petunia corollas. *Physiol. Plant*, 1989, 81, 127-133.

Weiss, M. R. Floral colour changes as cues for pollinators. *Nature*, 1991, 354, 227-229.

Weiss, M. R. Floral color change: a widespread functional convergence. *Am. J. Bot.* 1995, 82, 167-185.

Weyerstahl, P., Marschall, H., Bork, W. R. and Wilk, R. Megastigmanes and other constituents of the absolute of Boronia megastigma from Tasmania. *Liebigs Ann. Chem.*, 1994, 1043-1047.

Weyerstahl, P., Marschall, H., Bork, W. R., Rilk, R., Schneider, S. and Wahlburg, H. C. Constituents of the absolute of Boronia megastigma Nees from Tasmania. *Flav. Fragr. J.*, 1995, 10, 297-311.

Willstätter, R. and Mallison, H. Untersuchungen über die Anthocyane. X. Über Variationen der Blütenfarben. *Justus Liebigs Ann. Chem.* 1915, 408, 147-162.

In: Flowers
Editors: Teodor Berntsen and Kaj Alsvik

ISBN: 978-1-62808-798-7
© 2013 Nova Science Publishers, Inc.

Chapter 4

BEYOND THE PHYSIOLOGICAL ROLE IN PLANTS: FLOWERS AS SOURCES OF THERAPEUTIC MOLECULES

*Mahomoodally Mohamad Fawzi**

Department of Health Sciences, Faculty of Science,
University of Mauritius, Réduit, Mauritius

ABSTRACT

Innovation, sustainability and safety of drugs have become the main foci of the modern pharmaceutical industry. Consumer awareness of the possible side effects of the use of chemical-based therapeutic agents has compelled researchers to probe and explore natural botanical-based agents that are toxicologically safe, especially when used in the health care system. To this effect, alternative sources of drugs are being probed for novel lead molecules in an endeavour to offer permanent and sustainable solutions to patients. Plants with potential therapeutic value have been used since time immemorial to cure various ailments and it is only during the last past decades that the scientific community has expressed renewed interest in plants as alternative therapeutic sources of lead molecules. Flowers have traditionally been used in cooking in various cultures, such as European, Asian, East Indian, Victorian English, and Middle Eastern. Being an important part of a plant, flower as any botanical source possess secondary metabolites or the bioactive compounds (phytochemicals) which have been reported to be accountable for various observed biological activities and health benefits.

Phytochemicals produced from flowers exhibit pharmacological effects on the body ranging from anti-inflammatory and antimicrobial properties to cardiovascular benefits. This book chapter focuses on providing and updating available information on the therapeutic activities exhibited by common flowers (Saffron, Lavender, Chamomile, Citrus and Calendula), which are envisaged to find potential applications as alternative natural preservatives for foods and/or applications in the pharmaceutical industries to develop new and sustainable botanical-based products for treating/managing various human ailments.

---

* E-mail address: f.mahomoodally@uom.ac.mu; mmfawzi@gmail.com.

INTRODUCTION

In the past few years, there has been an increasing importance in industry, academia and the health sciences in medicinal and therapeutic properties of flowers. Flowers have traditionally been used in cooking in various cultures, such as European, Asian, East Indian, Victorian English, and Middle Eastern. Edible flowers can be used fresh as a garnish or as an integral part of a dish, such as a salad. Some flowers can be stuffed or used in stir-fry dishes (Kaisoon et al., 2012; Belsinger, 1991). In some Asian countries, many flowers have been eaten since ancient times, and their medicinal properties as well as nutritional value have been reported. For instance, a vast number of flowers are consumed in Thailand such as neem flower (*Azadirachia indica*), sesbania flower (*Sesbania grandiflora*), cassia flower (*Cassia siamea*), roses (*Rosa damascena*), marigolds (*Tagetes erecta*) and West Indian jasmine (*Ixora chinensis*). Flowers are used in salad, light curry or are used as vegetables (Krasaekoopt and Kongkarnchanatip, 2005; Kaisoon et al., 2012).

Since time immemorial, medical practitioners have acknowledged the therapeutic properties of certain flowers. More than just spanning time, this knowledge also extends across many cultures around the world. For instance, the local communities in Thailand believe that these edible flowers could cure illness and diseases, such as: diarrhea, stomach ache, nausea amongst others, which indicates their potential antimicrobial activity (Kaisoon et al., 2012; Belsinger, 1991). It is believed that one of the greatest advantages is that flowers and related products offer completely natural medicinal properties, often without the significant side effects that modern drugs and medications bring on. Additionally, natural remedies made from flowers can be much cheaper than drugs marketed by pharmaceutical companies (Tugba et al., 2012).

Historically, medicinal properties of flowers were discovered thousands of years ago. Folk medicines practiced by the natives around the globe extensively use flowers. Flower therapy uses essential flower oils, flower waters, flower juice, flower petals (fresh and dried), and aroma to heal both physical and spiritual components of the body. Because of growing epidemiologic evidence of the medicinal properties of flowers, modern medicines use flower products. Essential oils from flower are widely used in cosmetics. For instance, flowers of lavender, marigold, rose, passionflower and rosemary are widely exploited in aromatherapy. Additionally, rose, passionflower and chamomile are edible food products in some cultures and have been documented to harbor important therapeutic molecules that are used to manage and/or treat panoply of pathologies with minimum side effects (Tugba et al., 2012; Imelouane et al., 2009; Kaisoon et al., 2012).

SAFFRON - *CROCUS SATIVUS*

Saffron flower is one of the highly prized spices known since antiquity for its color, flavor and medicinal properties. It is the dried "stigma" or threads of the flower of the *Crocus sativus* plant. The plant is a bulbous perennial plant that belongs to the botanical family of Iridaceae and of the genus, *Crocus*. The distinct flavor of sarfron is due to its chemical composition which is mainly picrorocin, and safranal. It also contains a natural carotenoid chemical compound, crocin, which gives saffron its golden-yellow hue.

There is a long history of the use of saffron flowers in the traditional medicines of many cultures. Interestingly, during the past few years there has been increasing interest in the biological effects of saffron and its potential medical applications. To a certain extent, the recent interest in saffron is part of a generally increasing awareness of the great medical potential of natural products, particularly food plants and spices with low toxicity (Negbi, 1999).

In Chinese traditional medicine, saffron has been widely used for its anodyne, tranquilizing and emetic properties. It has also been used in the treatment/management of menstrual disturbances, cardiovascular diseases and some other diseases related to high blood viscosity. It has found applications in nervous disorders: *per se* to alleviate fears, cure trances and in the treatment of some disorders of the central nervous system (Zhou et al., 1987, Liakopoulou-Kyriakides and Skubas 1990). Its medical value was recorded in Yi Lin-Ji-Yao, a traditional Chinese medical book composed during the Ming Dynasty (16th century); notable among the effects described was the promotion of blood circulation to remove blood stasis. The book Yinshanzhengyao ("The Importance of Diet") contains 136 recipes which include saffron for treating a variety of conditions. Saffron also appears in several traditional Chinese pharmaceutical compendia (Ni 1992). It has been used in traditional Indian and Azerbaijani medicine to treat various diseases including cancer, heart disease, eye disease, blood disease and muscle paralysis (Damirov et al., 1988; Negbi, 1999).

As a result of a variety of recent scientific investigations, there is now convincing clinical evidence for the therapeutic activity of saffron and its constituents. One of the activities of saffron which has the greatest potential medical applicability is its ability to inhibit carcinogenesis. A number of recent studies have shown that saffron extract possesses antitumor activity against transplanted tumors and anti-carcinogenic activity against chemically induced carcino-genesis *in vivo,* and cytotoxic effects on tumor-derived cells *in vitro.* These findings have raised the possibility that natural saffron and/or some of its phytochemicals might be used as alternative antitumor or anti-carcinogenic agents, either alone or in combination with synthetic substances having anticancer activity. The recent scientific findings on the biological activities of saffron, together with the body of anecdotal evidence for its therapeutic activity against a number of diseases, provide strong indications that saffron and/or its components may be useful agents in modern medicine (Negbi, 1999).

Several mechanisms have been proposed for the antitumor effect of the carotenoid constituents of saffron. The observation (Nair et al., 1991a,b, 1992, 1994) that the antitumor effect of the extract could be demonstrated only when the drug was given orally but not when given intra-peritoneally led to the hypothesis that prior metabolism of the active component/s may be required for its/their antitumor activity. Specifically, crocin is suggested to exert its antitumor effect via its metabolic conversion to a retinoid. A second proposed mechanism for the antitumor action of carotenoids is based upon the widely accepted hypothesis that these phyto-compounds behave as antioxidants and thus function as inhibitors of free radicals generated progress of diseases (Bruce 1983, Burton and Ingold 1984). Most carotenoids are lipid-soluble and thus might be expected to act as membrane associated high-efficiency free-radical scavengers (Burton and Ingold 1984). This mechanism, involving the radical-trapping potential of carotenoids, has received support from computational molecular modeling studies (Neidle and Jenkins 1991, Martin 1991). A third mechanism involves the interaction of carotenoids with topoisomerase II, an enzyme involved in cellular DNA replication (Morjani et al., 1993). This idea is supported by the nuclear localization of some carotenoids (Manfait et al., 1991), as well as by their inhibitory effects on cellular DNA synthesis. A fourth suggested mechanism is that the cytotoxic effect of crocin is mediated via apoptosis (Wyllie 1992).

In addition saffron has been documented to possess plethora of medicinal properties as summarised hereunder:

- Saffron contains many phytochemicals that are known to have been antioxidant potential, disease preventing and health promoting properties.
- The flower stigma are composed of many essential volatile oils but the most important being safranal, which gives saffron its distinct hay-like flavor. Other volatile oils in saffron are cineole, phenethenol, pinene, borneol, geraniol, limonene, p-cymene, linalool, terpinen-4-oil, etc.
- Saffron flower has many non-volatile active components; the most important of them is α-crocin, a carotenoid compound, which gives the stigmas their characteristic golden-yellow color. It also contains other carotenoids, including zea-xanthin, lycopene, α- and β-carotenes. These are important antioxidants that help protect the human body from oxidative stress, cancers, infections and acts as immunomodulators.
- The active components in saffron have many therapeutic applications in many traditional medicines as antiseptic, antidepressant, anti-oxidant, digestive, anti-convulsant.
- This novel spice is a good source of minerals like copper, potassium, calcium, manganese, iron, selenium, zinc and magnesium. Potassium is an important component of cell and body fluids that helps control heart rate and blood pressure. Manganese and copper are used by the body as co-factors for the antioxidant enzyme, superoxide dismutase. Iron is essential for red blood cell production and as a co-factor for cytochrome oxidases enzymes.
- Additionally, it is also rich in many vital vitamins, including vitamin A, folic acid, riboflavin, niacin, vitamin-C that is essential for optimum health (Negbi, 1999).

LAVENDER - *LAVANDULA* SPECIES

Flowers of *Lavandula* species are mainly exploited for their essential oils, which are widely used in perfumery, cosmetics, food processing and nowadays also in aromatherapy products. The dried flowers have also been used since time immemorial in pillows, sachets amongst others for promoting sleep and a feeling of relaxation. Numerous lavender plants are also sold as ornamental plants for the garden; these include *L. latifolia, L. pinnata, L. lanata, L. dentata* and *L. stoechas* and their numerous cultivars (Imelouane et al., 2009; Hart and Lis-Balchin 2002)

The Abbess Hildegarde (1098-1179) wrote about lavender flowers and its ability to delouse and to clear eyes and also drive away evil spirits. The preservative properties of aromatic and medicinal plant volatile essential oils and extracts have been recognised since Biblical times, while attempts to characterise these properties in the laboratory date back to the 1900s (Deans 2002).

Lavender was used in many medicines in medieval Wales and England in conjunction with numerous other herbs. Gerard (1636) mentions spike lavender as being effective against catalepsy, migraine and fainting.

Psychological healing was also possible, for example, for passions of the heart and inward and outward grief. He mentioned the usage of lavender for the stomach, liver and obstructions in the spleen, for the kidneys and against colic and wind (suggesting a relaxant effect). He, however, also stated that it brought on menses and caused abortion of the dead child and after-birth (i.e. a contractile effect). Culpeper stated that true sweet lavender (*L. angustifolia*) was not considered good for anything medicinal, only as a perfume, and, he also warned against excessive use of the 'oil of spike' However, these statements have been forgotten or completely ignored and all the lavenders are now considered equal, in the absence of scientific proof (Hart and Lis-Balchin 2002).

Essential oils prepared from *L. angustifolia, L. dentata* and *L. spica* have been documented to be spasmolytic on intestinal smooth muscle and this activity is probably due to the linalool and linalyl acetate. The mechanism of action is myogenic involving cAMP at low concentrations with block of calcium channels also being involved at high concentrations. When there is also some spasmogenic activity this is probably due to the presence of α- and β-pinene and 1,8-cineole. Linalool has been shown to have a local anaesthetic action on motor nerve skeletal muscle preparations and it is surprising that a similar activity is not seen with the field stimulated guinea-pig ileum preparation (Imelouane et al., 2009).

In the whole animal, the oils of lavender cause sedation which may be explained by the observation that linalool is capable of displacing glutamate from its receptor. Lavender oil also depresses skeletal muscle activity in the rat but the work of Buchbauer *et al.* (1993) suggests that this is not important in humans. The fact that some extracts of *L. angustifolia* have a strong spasmogenic action in the dried flowers and fresh leaves is somewhat disturbing as so many modern herbal and aromatherapy books state that the teas are sedative and are often prescribed for upset stomachs. The results suggest that all the information has been mistakenly transcribed from early herbals where *L. spica*, a more camphoric lavender was used medicinally and not the very floral, *L. angustifolia*, which has always been used, as it is used today, mainly in perfumery and cosmetic products (except by aromatherapists who have no knowledge of plant taxonomy or past herbal literature). The spasmolytic results shown for the water soluble extracts of the more camphoraceous *L. stoechas* suggests that this mistake is probably the reason for the differences in the well-quoted action of the camphoraceous spike lavender and should not be confused with that of the non-canmphoraceous *L. angustifolia* (Lis-Balchin et al., 2002).

There is a long tradition in folklore of its use by ordinary people of lavender to affect psychological states. Nonetheless, scientific research into the psychological effects of lavender is limited. However, there is a long history of it being regarded, and used, as a sedative or calming agent. The effects on cells and brain tissues also suggest both reduction in electrical activity and in anticonvulsant effects. However, a review of research on the effects of lavender on EEG and psychological responses, both laboratory and clinically-based, reveals that responses to lavender may be determined, not only by these pharmacological sedative effects, but by individual, situational and expectational factors independent of the lavender odour itself. As a result, although experimental research indicates calming or sedative effects, these can be difficult to identify or pin down, especially in real-life applications (Michael Kirk-Smith, 2002).

CHAMOMILE - (*MATRICARIA CHAMOMILLA* L. SYN. CHAMOMILLA RECUTITA)

Chamomile flowers extracts belong to those drugs that experienced a wide medical application in ancient times. The curative effect of chamomile has been known by physicians for about 2500 years. Hippocrates gives a description of the drug in the 5th century B.C., and chamomile appears as a medicinal plant in the work *De Materia Medica* written by Dioscorides (1st century A. D.). Chamomile has been known for centuries and is well established in phytotherapy. In traditional folk medicine it is found in the form of chamomile tea, which is drunk internally in cases of painful gastric and intestinal complaints connected with convulsions such as diarrhea and flatulence, but also with inflammatory gastric and intestinal diseases such as gastritis and enteritis (Franke and Schilcher, 2005).

The yellow and white heads of the chamomile flower are used to make teas, extracts and ointments. These are used to reduce swelling; to inhibit bacterial, viral and fungal growth; and to treat nausea, vomiting and other digestive disorders. People with asthma should not use chamomile because it can worsen their symptoms. Chamomile preparations are mainly used because of their antiphlogistic, spasmolytic, and carminative activity. However, their bacteriostatic and fungistatic properties should not be underestimated. Application fields include dermatology, stomatology, otolaryngology, internal medicine, in particular gastroenterology, pulmology, pediatry, and radiotherapy.

The therapeutic effectiveness is in total due to the combined pharmacological and biochemical effects of several chamomile phyto-constituents. For therapeutic success, it is important to use standardized total extracts or the essential oil. It is suggested that the full spectrum of activity will not be reached by applying individual chamomile substances only (Schilcher *et al.,* 2005).

Externally chamomile is applied in the form of hot compresses to badly healing wounds, such as for a hip bath with abscesses, furuncles, hemorrhoids, and female diseases; as a rinse of the mouth with inflammations of the oral cavity and the cavity of the pharynx; as chamomile steam inhalation for the treatment of acne vulgaris and for the inhalation with nasal catarrhs and bronchitis; and as an additive to baby baths. In Roman countries it is quite common to use chamomile tea even in restaurants or bars and finally even in the form of a concentrated espresso. This is also a good way of fighting against an upset stomach due to a sumptuous meal, plenty of alcohol, or nicotine. In this case it is not easy to draw a line and find out where the limit to luxury is (Franke and Schilcher, 2005).

The most important components are chamazulene with an undisputed antiphlogistic effect as well as α-bisabolol, a sesquiterpene alcohol with antiphlogistic, antibacterial, antimycotic, ulcusprotective, and musculotropical-spasmolytical properties. Both of them are main constituents of the essential oil. The chamomile flavones have a musculotropical, a spasmolytical, and an antiphlogistic effect if locally applied. Further important active principles are the bisabolol oxides A and B, cis- and trans-spiroether, coumarins, and mucilage (Schilcher *et al.,* 2005; Franke and Schilcher, 2005).

The therapeutic value of chamomile preparations is not just based on one main active principle but on the combination of many individual ones (their effects were proved both clinically and by animal experiments and combine a total effect suitable for a wide range of indications). A chamomile extract containing all components in an optimum concentration should be given preference to a usual domestic infusion. When diluting such an extract with hot water the lipophile constituents are also partly incorporated in the tea, being most suitable both for internal as well as for external application. In addition the concentrated form of the chamomile extract makes use exceeding the usual chamomile therapy possible. The proven antiphlogistic effect is therapeutically used by various specialized disciplines on a broad basis and in different ways. A particularly favorable judgment should be passed on the fact that the corresponding preparations can be applied both internally and externally (Franke and Schilcher, 2005).

ORANGE, LEMON, CITRON - CITRUS SPECIES

The importance of flowers of some species of *Citrus* (orange, lemon, citron) in therapy and pharmacy received official recognition with the appearance of the first pharmacopoeias. Essential oils extracted from citrus flowers are largely employed as aromatizers in the food and pharmaceutical industries. In particular, lemon and bergamot essential oils are used in the cosmetic industry, for the production of perfumes, detergents and body-care products, mainly because of their fragrance and solvent properties. Besides, they are present in various pharmaceutical preparations for gynaecological, ophthalmic and surgical use, as also in dentistry, because of their well-known antiseptic properties. Apart from being used as disinfectants, owing to their powerful antimicrobial properties, recent studies point out the possibility of employing citrus essential oils and/or their active principles in clinical fields (also due to their wide range of activity and their reduced resistance phenomena), but also as preservatives in the food industry and as alternative pesticides in integrated programmes (Dugo and Giacomo, 2002).

Citrus essential oils are rarely utilized, as such, in the pharmacotherapeutic field. However, much has been achieved as regards knowledge of the biological properties of the active principles isolated from these essential oils. As a result, some of these active principles (or compounds derived from them) are being successfully employed in therapy (for example, 5-methoxypsoralen in the treatment of psoriasis and vitiligo); the possibility is also being closely examined of utilizing them in the prevention of certain pathological conditions (for example, d-limonene in the chemoprevention of tumoral disease). Furthermore, extensive toxicological studies have proved to be fundamental in regulating, on the basis of scientific criteria, the use of citrus essential oils and/or their active principles not only in the pharmacotherapeutic field, but also in quite different fields (such as that of the cosmetic industry and in de-fatting products) (Kharebeva and Tsertsvadze, 1986; Dugo and Giacomo, 2002).

Blossom of *Citrus aurantium*, commonly known as sour orange or bitter orange has recently been clinically studied for its anxiolytic effect and it was concluded that the flowers may be effective in terms of reduction in preoperative anxiety before minor operation (Akhlaghi et al., 2011). Traditionally, *Citrus aurantium* is used as an alternative medicine in some countries to treat anxiety, insomnia and as an anticonvulsant, suggesting depressive action upon the central nervous system (CNS). In an anxiety model study, *Citrus aurantium* was able to enhance the sleeping time induced by barbiturates. In animal model, this sedative effect was in accordance with traditional use of *Citrus aurantium* (Carvalho-Freitas and Costa et al., 2002; Akhlaghi et al., 2011).

MARIGOLD - CALENDULA SPECIES

Calendula L. (Asteraceae), also known as Garden Marigold, God-Bloom, Holligold, Marigold, Marybud, and Pot Marigold, is part of the botanical family of Asteraceae/Compositae (Jellin et al., 2003; Re et al., 2009). Marigold is a reputed medicinal plant with ornamental properties. The yellow or orange-colored flowers are used as food dye, spice, and tea as well as tincture, ointment or cosmetic cream. Recently, *C. officinalis* L. has become quite important in phytotherapy due to its healing effects against dermatological diseases (Bedi and Shenefelt, 2002; Leach, 2008; Fronza et al., 2009). The plant has been reported to contain mainly carotenoids, flavonoids, phenolic acids, and triterpenes (Kishimoto et al., 2005). The flower *per se*, flower extracts, flower essential oil, and seed oil of *C. officinalis* are cosmetic ingredients and the plant has been presented as a new source for cosmetic industry (Avramova et al., 1988). *C. officinalis* have been found to be usually safe (Andersen et al., 2010), although very rare occurrence of contact dermatitis due to Compositae (Asteraceae) plants should be taken into account (Paulsen, 2002). On the other hand, marigold was shown to be a promising dye plant for obtaining natural colors (Piccaglia and Venturi, 1998; Guinot et al., 2008).

According to ancient records (Keville, 1991), the Calendula flowers were used as a symbol of remembrance and believed to gives great forces of warmth and benign compassion to the human soul, especially helping to balance the active and receptive modes of communication. Besides, the Calendula essential oil has been reported to be used in care of the elderly (Buckle, 2003). Recently, investigators examined the inhibitory activity of the n-hexane, dichloromethane, acetone, ethyl acetate, methanol, and water extracts of the leaf and flowers of *Calendula arvensis* and *C. officinalis* L. against acetylcholinesterase (AChE) and butyrylcholinesterase (BChE), which are the key enzymes for the treatment of Alzheimer's disease and, lately, Down syndrome (Giacobini, 2004; Nieoullon, 2010). Since neurodegeneration is strongly associated with oxidative damage (Mariani et al., 2005), antioxidant activity of the extracts was tested by 2,2-diphenyl-1-picrylhydrazyl (DPPH), ferric ion chelating capacity, and ferric-reducing antioxidant power (FRAP) assays at 250, 500, and 1000μg/mL. Total phenol and flavonoids contents of the extracts were calculated spectrophotometrically (Andersen et al., 2010).

The leaf and flower extracts of *C. arvensis* and *C. officinalis* were examined for their antioxidant and inhibitory activity against AChE and BChE. Recent findings revealed that the methanol and ethyl acetate extracts obtained from the flowers of *C. arvensis* with higher DPPH radical scavenging activity and FRAP as well as AChE inhibitory activity might deserve a further evaluation to elucidate the active components, which are possibly flavonoid derivatives for antioxidant activity and triterpene derivatives for AChE inhibition.

The dried flower Calendula is also used as a spice and is considered to be generally recognized to be safe (GRAS) by the food and drug administration (Food and Drug Administration, 2007) and the Flavors and Extracts Manufacturers Association (FEMA) (FEMA number 2658). It is used topically as a natural anti-inflammatory medicine and for poorly healing wounds and leg ulcers. The dosages cited are 2-4 mL of tincture diluted to 250–500 mL with water or 2–5 g of herb in 100 g of ointment (Jellin et al., 2003). Other topical uses include treatment for 1st degree burns and scalds, bruises, boils, and rashes (Lueng and Foster, 1996; Blumenthal et al., 2000). A tea made from 1 to 2 g of the flower in 150 mL of boiling water has also been used up to three times a day as an antispasmodic (Jellin et al., 2003). Other oral uses include alleviation of the discomfort associated with stomach ulcers and inflammation of the oral and pharyngeal mucosa (Lueng and Foster, 1996; Blumenthal et al., 2000). Calendula preparations are therefore regarded as traditionally used medicines by the European Medicines Agency (EMEA) (EMEA, 2008a,b). In cosmetic or personal care preparations calendula extracts are used as skin conditioning agents at concentrations ranging up to 1% but are generally below 0.1%. All together, these data allow one to conclude that calendula extracts should be considered as having a long history of safe use as defined by Constable et al. (2007).

CONCLUSION

Flowers may represent a large source of chemicals which can be of great therapeutic value. The widespread practice of traditional medicines in developing countries as well as in developed countries together with the rapidly increasing demand for alternative medicinal remedies largely encourages research and development in this field. Further motivation is brought about by the Beijing Declaration published by the World Health Organization in 2008 which calls for an integration of traditional remedies in modern medicine. The outcomes of conventional medicine are not always satisfactory in the treatment of various diseases. Due to numerous side-effects associated with conventional medicine, the fear of using corticosteroids as a remedy and overall failure of conventional remedies contribute to such patients' complaints. This is why large number of people are now turning towards traditional remedies. Medicinal flowers are highly expected to yield fruitful results in the discovery of new therapeutic molecules, mainly those to be used against resistant microorganisms, in inflammatory and allergic skin disorders and as cosmetics for the wellness industry. An advanced investigation on the less commonly used medicinal flowers plants with the aim of identifying any novel molecule(s) should be part of future research.

REFERENCES

Akhlaghi M, Shabanian G, Rafieian-Kopaei M, Parvin N, Saadat M, Akhlaghi M. *Citrus aurantium* blossom and preoperative anxiety. *Rev. Bras. Anestesiol.,* 2011; 61: 6: 702-712.

Andersen FA, Bergfeld WF, Belsito DV, Hill RA, Klaassen CD, Liebler DC, Marks Jr JG, Shank RC, Slaga TJ, Snyder PW. Final report of the cosmetic ingredient review expert panel amended safety assessment of *Calendula officinalis*-derived cosmetic ingredients. *Int. J. Toxicol.* 2010; 29: 221S-243S.

Avramova S, Portarska F, Apostolova B, Petkova S, Konteva M, Tsekova M, Kapitanova T, Maneva K. Marigold (*Calendula officinalis* L.): Source of new products for the cosmetic industry. *Medico-Biol. Inform.* 1988; 4: 24-26.

Bedi MK, Shenefelt PD. Herbal therapy in dermatology. *Arch. Dermatol.* 2002; 138: 232–242.

Belsinger S. *Flowers in the kitchen.* Loveland, Colorado: Interweave Press. 1991.

Blumenthal M, Goldberg A, Brinckmann J. Calendula flower. In: Herbal Medicine Expanded Commission E Monographs. *American Botanical Council*, Austin, 2000; 4-45.

Bruce NA. Dietary carcinogens and anticarcinogens. Oxygen radicals and degenerative diseases. *Science.* 1983; 221: 1256-1264.

Buchbauer G, Jirovetz L, Jäger W, Dietrich H, Plank C, Karamat E. Romatherapy: evidence for sedative effects of the essential oil of lavander after inhalation. *Z Naturforsch C.* 1991; 46: 1067-1072.

Buckle J. Care of the elderly-essential oils in practice. In: *Clinical Aromatherapy*, 2nd edition. Elsevier Ltd., Amsterdam, The Netherlands. 2003.

Burton GW, Ingold KU. ß-carotene: An unusual type of lipid antioxidant. *Science.* 1984; 224: 569-573.

Carvalho-Freitas MI, Costa M. Anxiolytic and sedative effects of extracts and essential oil from *Citrus aurantium* L. *Biol. Pharm. Bull.*, 2002; 25: 1629-1633.

Constable A, Jonas D, Cockburn A, Davi A, Edwards G, Hepburn P, Herouet- Guicheney C, Knowles M, Moseley B, Oberdörfer R, Samuels F.History of safe use as applied to the safety assessment of novel foods and foods derived from genetically modified organisms. *Food Chem. Toxicol.* 2007; 45: 2513–2525.

Damirov IA, Prilipko LI, Shukurov DZ, Kerimov JB. Remedy Plants of Azerbaijan. *Journal.* 1988; 90-93.

Dugo G, Di Giacomo A. *Medicinal and Aromatic Plants - Industrial Profiles.* Taylor and Francis. 2002.

EMEA, 2008a. European Medicines Agency, Committee on Herbal Medicinal Products (HMPC). *Assessment Report for the Development of Community Monographs and for Inclusion of Herbal Substance(s),* Perparation(s) or Combinations Thereof in the Community List. EMEA/HMPC/179282/2007, London 6 March 2008. <www.emea.europa.eu/pdfs/human/hmpc/ *calendula_officinalis_*flos/17928207en.pdf>.

EMEA, 2008b. European Medicines Agency, Committee on Herbal Medicinal Products (HMPC). Community Herbal Monograph on *Calendula Officinalis* L. Generally Recognized as Safe 21 CFR Part 182.10. Flos. EMEA/HMPC/179281/2007Coor, London, 2 May 2008. <http://www.emea.europa.eu/pdfs/human/hmpc/*calendula_officinalis_*flos/17928107enfin.pdf>.

Food and Drug Administration, 2007. *Food and Drug Administration Department of Health and Human Services.* Food for Human Consumption; Substances.

Franke R, Schilcher H. Chamomile Industrial Profiles. *Medicinal and Aromatic Plants - Industrial Profiles.* CRC Press Taylor & Francis Group. 2005.

Fronza M, Heinzmann B, Hamburger M, Laufer S, Merfort I. Determination of the wound healing effect of Calendula extracts using the scratch assay with 3T3 fibroblasts. *J. Ethnopharmacol.* 2009; 126: 463-467.

Giacobini E. Cholinesterase inhibitors: new roles and therapeutic alternatives. *Pharmacol. Res.* 2004; 50: 433-440.

Guinot P, Gargadennec A, Valette G, Fruchier A, Andary C. Primary flavonoids in marigold dye: extraction, structure and involvement in the dyeing process. *Phytochem. Anal.* 2008; 19: 46-51.

Imelouane B, Elbachiri A, Ankit M, Benzeid H, Khedid K. Physico-Chemical Compositions and Antimicrobial Activity of Essential Oil of Eastern Moroccan Lavandula dentata. *Int. J. Agri. Biol.* 2009; 11: 113-118.

Jellin JM, Gregory PJ, Batz F, Hitchens K, et al. 2003. P*harmacist's Letter Prescriber's Letter Natural Medicines Comprehensive Database,* 5th ed. Stockton, CA, pp. 265-66.

Kaisoon O, Konczak I, Siriamornpun S. Potential health enhancing properties of edible flowers from Thailand. *Food Research International* 46 (2012) 563-571.

Kharebeva, L.G. and Tsertsvadze, V.V. Volatile compounds of flowers of *Citrus unshi* Marc. *Subtrop. Kult.*, 1986, 1, 119-121.

Keville K. *The Illustrated Herb Encyclopedia*. Mallard Press, New York, USA. 1991.

Kirk-Smith M. The psychological effects of lavender. *In*: Lavender - The genus Lavandula. Maria Lis-Balchin. *Medicinal and aromatic plants - industrial profiles*. Harwood Academic Publishers imprint. 2002.

Kishimoto S, Maoka T, Sumitomo K, Ohmiya A. Analysis of carotenoid composition in petals of Calendula (*Calendula officinalis* L.). *Biosci. Biotechnol. Biochem.* 2005; 69: 2122-2128.

Krasaekoopt W, Kongkarnchanatip A. Anti-microbial properties of Thai traditional flower vegetable extracts. *AssumptionUniversity Journal of Technology*, 2005; 9(2), 71-74.

Leach MJ. *Calendula officinalis* and wound healing: a systematic review. *Wounds.* 2008; 20: 236–243.

Liakopoulou-Kyriakides M, Skubas AI. Characterization of the platelet aggregation inducer and inhibitor isolated from Crocus sativus. *Biochemistry International.* 1990; 22: 103–110.

Lueng AY, Foster S. Calendula. In: *Encyclopedia of Common Natural Ingredients used in Food Drugs and Cosmetics*. John Wiley & Sons, New York, 1996; 113-114.

Manfait M, Morjani H, Efremov R, Angibousi JF, Polissiou M, Nabiev F. High sensitive detection of intracellular carotenoids in single living cancer cells as probed by surface-enhanced Raman spectroscopy. In R.E.Hester and R.B.Giring (eds.), *Spectroscopy of Biological Molecules*, Royal Society of Chemistry, UK, 1991; 303-304.

Mariani E, Polidori MC, Cherubini A, Mecocci P. Oxidative stress in brain aging, neurodegenerative and vascular diseases: an overview. *J. Chromatogr. B.* 2005; 827: 65-75.

Martin YC. Computer-associated rational drug design. In: J.Langone (ed.), *Methods in Enzymology, Molecular Design and Modeling: Concepts and Applications, Vol. 203 Part B: Antibodies and Antigens, Nucleic Acids, Polysaccharides, and Drugs*, Academic Press, New York, pp. 587-613. 1991.

Morjani H, Riou JF, Nabiev Y, Lavlle F, Manfait M. Molecular and cellular interaction between intoplicine, DNA and topoisomerase II studied by surface-enhanced Raman scattering spectroscopy. *Cancer Res.* 1993; 53: 4784-4790.

Morjani H, Tantilis P, Polissiou M, Manfait M. Growth inhibition and induction of erythroid differentiation activity by crocin, dimethylcrocetin and b-carotene on K562 tumor cells. *Anticancer Res.* 1990; 10: 1398-1406.

Nair SC, Pannikar B, Panikkar KR. Antitumor activity of saffron (*Crocus sativus*). *Cancer Lett.* 1991a; 57: 109-114.

Nair SC, Salomi MJ, Panikkar B, Panikkar KR. Modulatory effects of the extracts of saffron and *Nigella sativa* against cisplatinum induced toxicity in mice. *J. Ethnopharmacol.* 1991b; 31: 75-83.

Nair SC, Salomi MJ, Varghese CD, Panikkar B, Panikkar KR. Effect of saffron on thymocyte proliferation, intracellular glutathione levels and its antitumor activity. *Bio Factors*. 1992; 4: 51-54.

Nair SC, Varghese CD, Panikkar KR, Kummboor SK, Parathod RK. Effects of saffron on vitamin A levels and its antitumor activity on growth of solid tumors in mice. *Int. J. Pharmacog*. 1994; 32:105-114.

Negbi M. *Medicinal and aromatic plants: industrial profiles*; SAFFRON *Crocus sativus* L. Harwood Academic Publishers imprint. 1999.

Neidle S, Jenkins TC. Molecular modeling to study DNA interaction by antitumor drugs. In J. Langone, (ed.), *Methods in Enzymology, Molecular Design and Modeling: Concepts and Applications, Vol. 203 Part B: Antibodies and Antigens, Nucleic acids, Polysaccharides, and Drugs*, Academic Press, New York, 1991; 203: 433-458.

Ni X. Research progress on the saffron (*Crocus sativus*). *Zhongcaoyao*. 1992; 23: 100-107.

Nieoullon A. Acetylcholinesterase inhibitors in Alzheimer's disease: further comments on their mechanisms of action and therapeutic consequences. *Psychol. Neuropsych. Vieil.* 2010; 8:123-131.

Paulsen E. Contact sensitization from Compositae-containing herbal remedies and cosmetics. *Contact Derm.* 2002; 47: 189-198.

Piccaglia R, Marotti M, Chiavari G, Gandini N. Effects of harvesting date and climate on the flavonoid and carotenoid contents of marigold (*Calendula officinalis* L.). *Flav. Fragr. J.* 1997; 12: 85-90.

Piccaglia R, Venturi G. Dye plants: a renewable source of natural colours. *Agro Food Ind. Hi-Tech*. 1998; 9: 27-30.

Re TA, Mooney D, Antignac E, Dufour E, Bark I, Srinivasan V, Nohynek G. Application of the threshold of toxicological concern approach for the safety evaluation of calendula flower (*Calendula officinalis*) petals and extracts used in cosmetic and personal care products. *Food Chem. Toxicol.* 2009; 47: 1246-1254.

Schilcher H, Imming P, Goeters S. Active chemical constituents of *Matricaria chamomilla* L. syn. Chamomilla recutita (L.) Rauschert. In. Rolf Franke and Heinz Schilcher. *Chamomile Industrial Profiles. Medicinal and Aromatic Plants - Industrial Profiles*. CRC Press Taylor & Francis Group. 2005.

Stanley GD. Antimicrobial properties of lavender volatile oil. In: Lavender - The genus Lavandula. Maria Lis-Balchin. *Medicinal and aromatic plants - industrial profiles.* Harwood Academic Publishers imprint. 2002.

Stephen H, Lis-Balchin M. Pharmacology of Lavandula essential oils and extracts in vitro and in vivo. In: Lavender - The genus Lavandula. Maria Lis-Balchin. *Medicinal and aromatic plants - industrial profiles.* Harwood Academic Publishers imprint. 2002.

Tugba E, Fatma SS, Ilkay EO, Gulnur T. Comparative assessment of antioxidant and cholinesterase inhibitory properties of the marigold extracts from *Calendula arvensis* L. and *Calendula officinalis* L. *Ind. Crop. Prod.* 2012; 36: 203-208.

Wyllie AH. Apoptosis and the regulation of cell numbers in normal and neoplastic tissues: an overview. *Cancer Metastasis Rev.* 1992; 11: 95-103.

Zhou Q, Sun Y, Zhang X. Saffron: *Crocus sativus* L. *J. Traditional Chinese Med.*1987; 28: 59-61.

In: Flowers
Editors: Teodor Berntsen and Kaj Alsvik

ISBN: 978-1-62808-798-7
© 2013 Nova Science Publishers, Inc.

Chapter 5

TO BE OR NOT TO BE A STAMINODE: THE FLORAL DEVELOPMENT OF *SAUVAGESIA* (OCHNACEAE) REVEALS DIFFERENT ORIGINS OF PRESUMED STAMINODES

J. Farrar and L. P. Ronse De Craene
Royal Botanic Garden Edinburgh, Scotland, UK

ABSTRACT

Sauvagesia erecta L. is a small pantropical herb in the family Ochnaceae. Flowers are unusual in that in addition to petals there is an inner whorl of five petaloid structures and clusters of small spathulate appendages situated between the petals and petaloid structures. The petaloid structures overlap to form a cone enclosing the androecium and gynoecium. Both the spathulate appendages and petaloid structures have traditionally been considered to be staminodial in origin.

The floral development and floral anatomy of *S. erecta* was investigated using scanning electron microscopy and light microscopy. Our analysis shows the late initiation of antepetalous petaloid structures following petals, gynoecium, and antesepalous stamens. This is consistent with the phyllotactic and developmental sequence for a second whorl of stamens. Further to this the vascular traces feeding both the petaloid structures and the stamens split from a common trace low in the receptacle of the flower. These observations confirm the staminodial nature of these petaloid structures. The primordia for the spathulate appendages arise very late in the ontogeny of the flower after all other organs have initiated and are already well developed. The receptacle between the petals and petaloid staminodes expands into a narrow androgynophore and the initial spathulate appendages arise in an antesepalous position below the stamens and staminodes. Further primordia develop around the first-formed appendages in centrifugal sequence, eventually joining neighbouring appendages in a continuous girdle. The vascular traces associated with the appendages are randomly attached to the staminal traces and divide repeatedly before connecting to individual structures.

The nature of a corona is discussed and it is concluded that the generally used definition is not adequately reflecting the homology of structures described as a corona. Our findings strongly suggest that the appendages are novel organs and could be considered as equating to the corona structures seen for example in Passifloraceae, Velloziaceae or Amaryllidaceae. Therefore the appendages are best interpreted as hypanthial emergences and not as staminodes. Comparison with other Ochnaceae suggests that outer filamentous staminodes found in several genera are best interpreted as a corona, in contrast to inner antepetalous staminodes. However, the function of the appendages in plant-insect interactions is unclear, except for attraction. Flowers of *Sauvagesia* are buzz-pollinated and the cone-like petaloid staminodes would appear to have a dual function in protecting the reproductive organs from damage and in directing the pollen released onto visitors.

INTRODUCTION

The family Ochnaceae consists of 27 genera in two subfamilies Ochnoideae and Sauvagesioideae occuring in tropical South America, Africa and Asia (Amaral, 1991). On the basis of molecular data two additional families Medusagynaceae (with one species) and Quiinaceae (with four genera) have been included as well (APG, 2009). Most genera in Ochnaceae have pentamerous flowers with a calyx of five sepals, five showy petals, stamens in multiples of five, and five carpels (occasionally two to three or more), which are fused to some extent and at least partially at the base, i.e. syncarpous. Stamens are generally arranged in two whorls (diplostemony) and one whorl can be sterile (staminodial) or absent (Dwyer, 1945; Amaral, 1991; Matthews et al., 2012). In a few genera flowers are polyandrous with fascicled stamens (e.g. *Ochna, Cespedesia, Luxemburgia*). A characteristic of Ochnaceae is that the pollen is shed from the anthers through short slits or more elaborate pores at the apex of the stamen, i.e. the stamens are poricidal (Kubitzki & Amaral, 1991). Poricidal stamens are a feature associated with "Buzz" pollination, in which the pollinating insects, generally bees, create a vibration in the stamens to release the dry pollen – an excellent description of the mechanism is found in Renner (1983). The stamens of Ochnaceae are large, generally with short basifixed filaments, and are easily knocked off.

Sauvagesia, a basal genus in Ochnaceae (Wurdack & Davis, 2006), has a pantropical distribution and is predominantly found in the neotropics, where it has diversified into approximately forty species (Zappi & Lucas, 2002). *Sauvagesia erecta* L., described in this chapter, is widespread with a range spanning Madagascar, tropical Africa and South and Central America, (Stevens et al., 2001 onwards). It is an herbaceous plant of savannahs and scrubby woodland (Iturralde, 2006) exhibiting small, showy pentamerous flowers (Figure 14) with a tripartite conical gynoecium. A common feature of the genus *Sauvagesia* are the five petaloid structures in an antepetalous position which overlap and form a cone surrounding both the fertile stamens and the gynoecium. These structures have generally been described as a staminodial whorl (e.g. Gilg, 1893; Dwyer, 1945; Amaral, 1991). In some species of *Sauvagesia* that have been described as the synonym *Lavradia,* the petaloid structures forming the cone are fused (Eichler, 1878; Dwyer, 1945; Zappi & Lucas, 2002). This cone, often referred to as a staminodial corona, effectively protects the androecium from damage caused by pollinators or potential pollen thieves and functions as a hose in the expulsion of pollen.

In addition to the five white to pink petaloid staminodes, *Sauvagesia erecta* has a girdle of small red spathulate to reniform appendages inserted between the petals and the stamen/staminode whorls. These spathulate appendages have been generally homologized with true staminodes (Engler, 1876; Gilg, 1893; Goebel, 1933: Dwyer, 1945; Cronquist, 1981; Amaral, 1991; Matthews et al., 2011). Dwyer (1945), in his monograph of the genus, describes both structures as an interior and exterior corona. Amaral (1991) distinguished petaloid staminodes against filiform staminodes ("fadenförmige Staminodien"). The presence of an outer whorl of spathulate staminodes in addition to the antepetalous whorl of petaloid staminodes would mean that *Sauvagesia* has two kinds of staminodes, which is a highly unusual situation in angiosperms. Contrary to previous authors, Eichler (1871, 1878) considered both kinds of structures as a corona ("paracorollinische Bildungen"). Ronse De Craene and Smets (2001) also questioned the evidence to interpret the outer spathulate appendages as staminodes, and considered them more akin to receptacular outgrowths or a corona. Contrary to Eichler, they considered the inner whorl to be staminodial.

There are few floral developmental studies in Ochnaceae, although more morphological evidence is needed to understand the characters shared by the large order Malpighiales (Wurdack & Davis, 2009; Endress & Matthews, 2012). A preliminary study, including *Sauvagesia erecta* and *S. glandulosa*, was carried out by Amaral and Bittrich (1998). We investigated the floral development and anatomy of *Sauvagesia erecta* to definitively determine the homology of the spathulate appendages observed in *Sauvagesia erecta* and to confirm the petaloid staminodes as being true staminodes.

RESULTS

The Ontogeny of the Major Floral Organs

The ontogeny of the major floral organs in *Sauvagesia erecta* was investigated with the Scanning Electron Microscope, using immature inflorescences[1]. Material was analysed over the full sequence of development from bract initiation (Figure 1) to mature, dehiscing flowers (Figure 14). Inflorescences consist of several lateral cymose partial inflorescences, which develop acropetally and could thus be described as thyrses. The partial inflorescences consist of paired flowers and are subtended by a bract and two bracteoles. The bract initiates before the two lateral bracteoles that emerge in sequence while the apical part of the stem differentiates as a terminal flower (Figure 1). A second lateral flower initiates later in the axil of one of the bracteoles (Figures 1, 3). The calyx arises in a spiral 2/5 sequence and shows a continuous, rapid development of all organs (Figures 1, 2). Sepals develop as broad, triangular organs with apical appendage and rapidly enclose the young buds (Figures 3, 4). The distinctive pentamerous shape of the upper bud in figures 1 and 2 is in part caused by the compression by the developing bracteoles and sepals. The five petal primordia initiate simultaneously on the rounded corners of the pentagon and are immediately followed by an antesepalous stamen whorl (Figure 5).

The development of the stamens is relatively rapid in comparison to that of the petals, which lag behind in their development (Figures 4-10). Petals develop as broadly triangular primordia with narrow base and only overtop the stamens in preanthetic buds. After initiation petals and stamens show a strong contortion and this is accompanied by the contorted aestivation of the petals at maturity (Figure 5). The stamens have short filaments and two,

---

[1] Buds (700 La, 703 La) were collected by LRDC in Belize (Program for Belize, Hillbank) and are kept at RBGE.

elongated rectangular thecae. On maturity the thecae dehisce via broad apical slits on the stomium of each theca (Figure 13). The stomium is visible as a slightly puckered line of cells running along the length of the groove on each theca (Figures 10-13). The apical dehiscence is thus the consequence of an incomplete lateral dehiscence of the anthers. More diverged genera of Ochnaceae, for example *Ouratea* and *Ochna*, have elaborate pores at the apex of the theca (Matthews et al. 2012). The simple poricidal theca of *Sauvagesia* is due to the presence of fibres in the endothecium running perpendicular to the exothecium; as the fibres dry they contract and split open the cells of the stomium (Stevens, 2001 onwards).

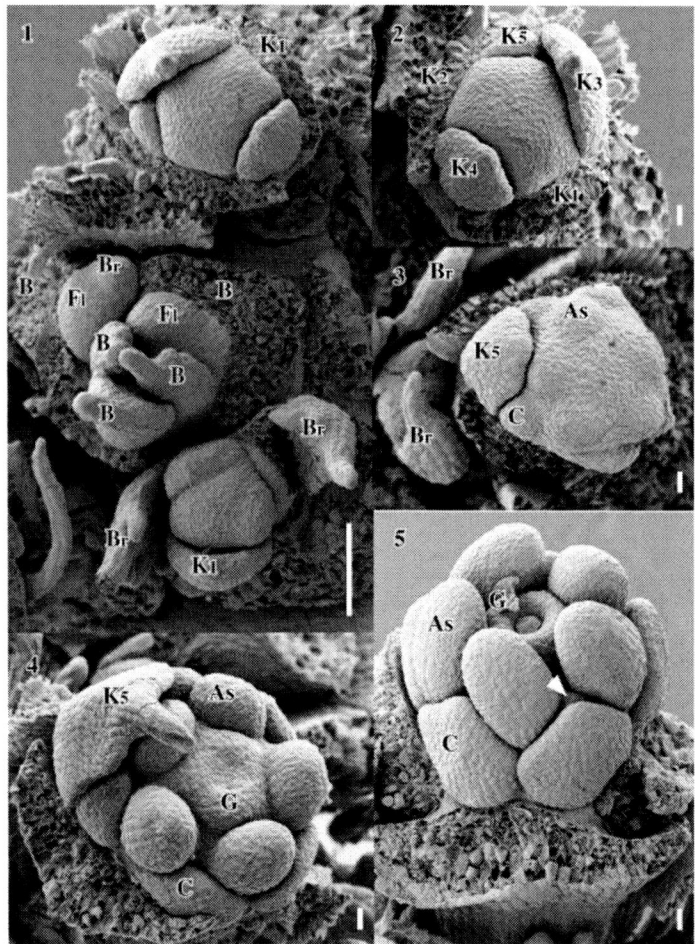

Abbreviations: As, antesepalous stamen; B, bract; Br, bracteole; C, petal; Fl, flower bud; G, gynoecial ring primordium; K, sepal. Numbers indicate order of initiation. Bars: 1=100μ, 2-5= 20μ.

Figures 1-5. Early stages in the floral initiation of *Sauvagesia erecta*.
1. Apical view of young inflorescence showing the sequential initiation of flowers in the axil of bract and bracteoles. Older flowers with sepal initiation. 2. Detail of top-flower in Figure 1 with sequential initiation of sepals and formation of pentagonal apex; outer sepals removed. 3. Flower with simultaneous initiation of corolla and antesepalous stamens. 4. Differentiation of petals and stamens and initiation of gynoecial ring primordium. 5. Initiation of antepetalous staminode (arrow). Note the extension of the gynoecial ring primordium and contortion of flower.

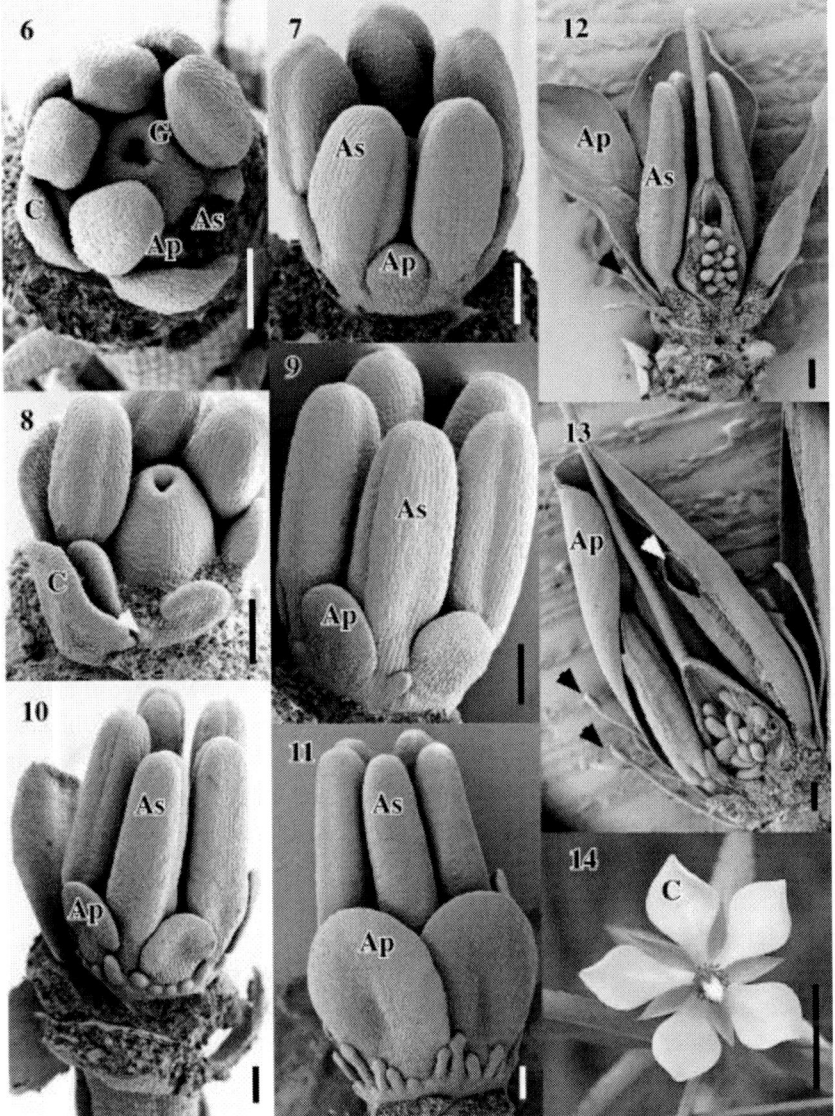

Abbreviations: As, antesepalous stamen; Ap, antepetalous staminode; C, petal; G, gynoecium. Bars: 1-11=100μ, 12-13= 200μ, 14= 500μ.

Figures 6-14. Mid- to late developmental stages in the floral initiation of *Sauvagesia erecta*.
6. Differentiation of antepetalous staminodes and conical gynoecial wall; one stamen removed. 7. Lateral view of flower at anther differentiation. Note the basal expansion of the receptacle below the androecium. 8. Lateral view of partly dissected androecium. Arrow points to initiation of appendage primordium on the receptacular extension below the stamen (removed). 9. Lateral view of older appendage primordium and expansion of staminodes. 10. Latero-centrifugal development of appendage primordia and extension of anthers. 11. Preanthetic bud showing three levels of complexity, with anthers, staminodes and crown of spathulate appendages. 12. Partly dissected flower at ovule differentiation; calyx and corolla removed. Arrow points to spathulate appendage. 13. Partly dissected flower at anthesis. Note the poricidal dehiscence of the anther (white arrow) and contorted staminodes surrounding stamens and style. Spathulate appendages shown with black arrows. 14. Mature flower showing three distinct petaloid areas: the white corolla, crown of red appendages and white staminodes.

In typical diplostemonous flowers the second whorl of stamens in antepetalous position would initiate immediately after the first antesepalous whorl. However, in the case of *S. erecta* the gynoecium initiates well before the emergence of the primordia of the second antepetalous whorl of stamens (Figures 4-5). The antepetalous whorl, which develops into the petaloid structures initiates very late at the time of contortion of the flower (Figure 5). The shape of the stamens - short filaments and long, thick thecae - is well established at the early stages of the petaloid staminode development (Figure 5). Staminodial primordia are initially globular but they become flattened and petal-like by marginal growth (Figures 6-9). Further petal and staminode development progresses swiftly with the petaloid staminodes developing characters similar to the petals (Figures 10-11). These characters include lobed bases in petals (Figure 7, 10) and similar surface cell types (J. Farrar, unpubl. data). In older buds staminodes also have a more pronounced stamen appearance by developing a narrow base (Figure 10) and a dorsal invagination (Figure 11).

The gynoecium arises as a ring primordium without indication of individual carpels (Figure 4); it is only through the development of septa that carpel number can be detected, which is generally three (Figure 12), rarely two. The gynoecial wall initially grows as a tubular structure around the rounded floral apex (Figure 5). Next the wall extends upwards and becomes triangular in shape (Figures 6, 8). A single long style develops on top of a cylindrical ovary (Figures 12, 13). The stigma consists of a single apical hole. Initiation of the septa was not investigated, but the ovary is trilocular in the lower two-thirds, where the ovules are positioned and unilocular in the upper one-third by divergence of the septa (Figure 12; cf. Matthews and Endress, 2012). Ovules are bitegmic and anatropous and are arranged in two irregular rows on each side of the septum wall.

Following the initiation of the staminodes there is an expansion of the receptacle between the petals and stamens to form a short hypanthium (Figures 7-11). This hypanthium could be interpreted as a small androgynophore as the receptacle between the petals and the entire reproductive region of the flower has been extended slightly. The stamens expand quickly during the early stages of development and reach a mature size of approximately double the length of the carpels (Figure 12). The petaloid staminodes extend past the top of the stamens (Figure 12) and on maturity are 1.5 times the lengths of the stamens. The petaloid staminodes curl round the stamens and carpel in a cone with a small aperture at the apex. The style protrudes from the aperture and well beyond the lip of the hole (Figure 13).

The initiation of the first primordia of the spathulate appendages occurs on the intervening hypanthium in an antesepalous position between the petaloid staminodes (Figures 8, 16 – marked with arrows). The primordia of the subsequent appendages (Figure 17 – marked with arrow) arise around the initial primordia in a centrifugal manner (Figures 18-21). The second tier consists of two primordia initiating basipetally left or right of the top primordium (Figures 9, 17, 18). A third tier is rapidly followed by a fourth, which connects the antesepalous groups in a continuous girdle (Figures 10, 11, 18-20). A fifth tier arises haphazardly on the outer limits of the girdle (Figures 20-21). The spathulate appendages rapidly elongate in a centrifugal sequence. Primordia are initially finger-like (Figure 19), but the apical part rapidly expands in a broad spathulate to reniform section (Figures 20-22). In preanthetic buds, the appendages reach halfway the petaloid staminodes and form a distinct cluster of bright red structures (Figures 12-14, 22).

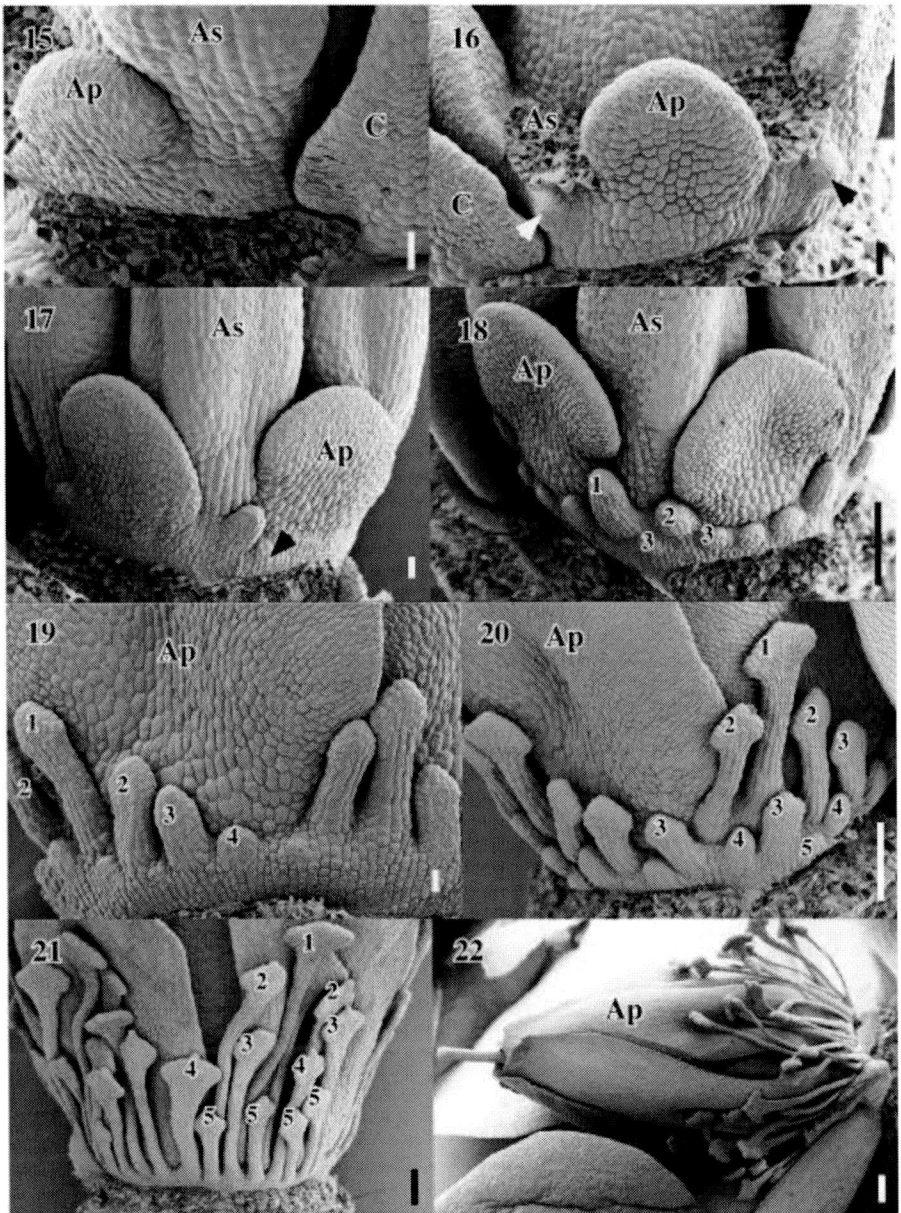

Abbreviations: As, antesepalous stamen; Ap, antepetalous staminode; C, petal. Numbers indicate
    sequence of appendage initiation. Bars: 15-17,19=20μ, 18,20,21=100μ, 22=200μ.

Figures 15-22. Development of the spathulate appendages in *Sauvagesia erecta*.
15. Lateral view of the base of staminodes and stamens prior to appendage initiation. 16. Initiation of
first appendage primordium opposite fertile stamen. 17. Expansion of stamens and staminodes and
initiation of second appendage primordium (arrow). 18. Initiation of third tier of primordia around first
formed appendages. 19. Initiation of fourth tier of appendages and merging of separate groups. 20.
Initiation of fifth tier and differentiation of triangular apex on first-formed appendages. 21. Preanthetic
bud showing expansion of the appendage crown with all appendages formed. 22. Lateral view of
preanthetic bud; perianth removed. Note the tubular staminodial whorl surrounding style and stamens,
as well as crown of spathulate appendages.

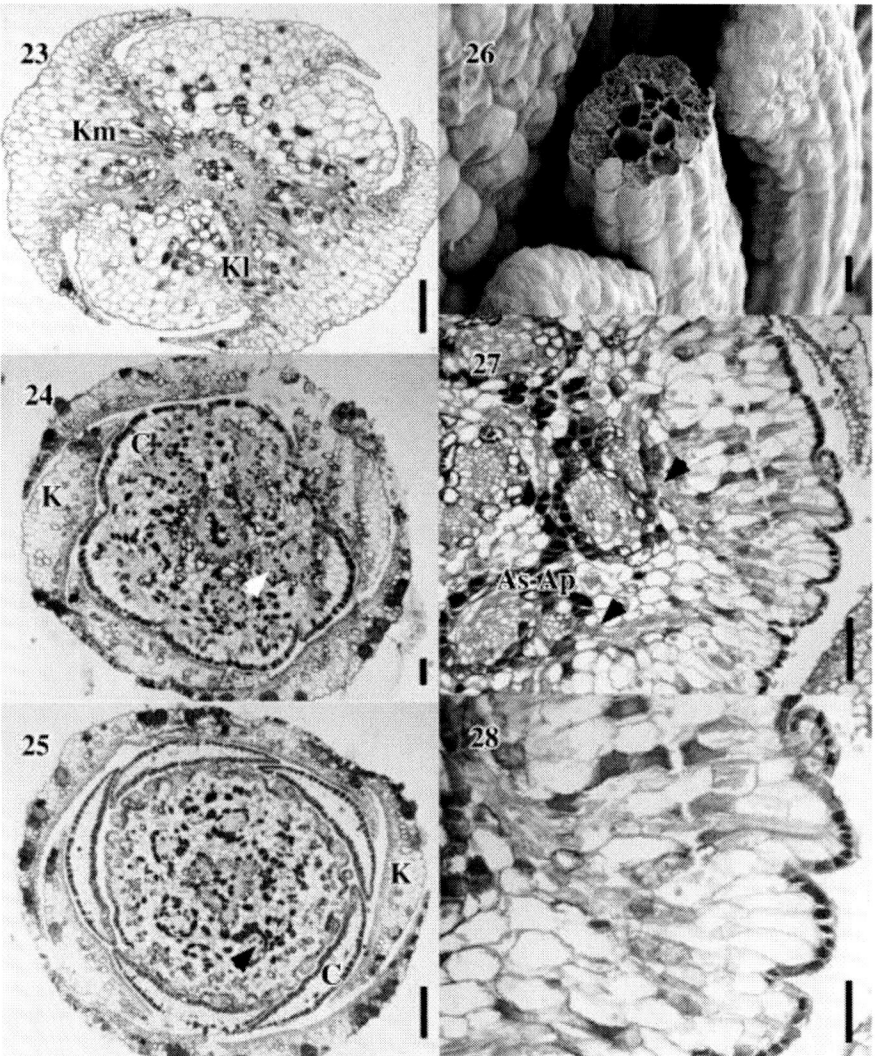

Abbreviations: As-Ap: common stamen-staminode bundle; C, petal trace or petal; K, sepal; Km, median sepal trace; Kl, commissural lateral sepal trace. Bars: 26=10µ, 28=50µ, 23,25=200µ, 24,27=100µ.

Figures 23-28. Floral anatomy of *Sauvagesia erecta* with special reference to the vasculature of the spathulate appendages.
23. Transverse section through the base of the flower showing the departure of sepal traces; commissural lateral traces precede the median traces. 24. Transverse section at the departure of the petal traces. White arrow points to the separation of a common stamen bundle from the petal-stamen bundle complex. Note that there are no bundles to the coronal appendages 25. Transverse section at the separation of common stamen-staminode bundles (arrow). The outline of the appendages is visible in the periphery, but no connecting traces are seen. 26. Transversally sectioned appendage showing the central vascular system. 27. Transverse section at the level of separation of stamen and staminode traces. Arrows point to departing appendage traces branching in the receptacle. 28. Detail of branching appendage trace and outline of the appendages at the periphery.

Vasculature of the Flower and the Spathulate Appendages

Transverse sections of the flower show the departure of commissural lateral bundles to the sepals diverging before the median sepal bundles (Figure 23). The outline of the sepals corresponds to their initiation in a 2/5 sequence. Petal bundles are detached from traces common with the androecium (Figure 24, white arrow). Petal traces immediately divide in several bundles before entering the base of the petals. Above the detachment of the petals, the single androecium bundles divide in the respective stamen and staminode bundles above the level where spathulate appendages are visible (Figure 25, arrow). The remaining stele is reorganized as three dorsal bundles alternating with three ventral bundles, which enter the base of the gynoecium at higher levels (not shown). When sectioned the spathulate appendages show a central vascular bundle consisting of a few xylem cells surrounded by an epidermis (Figure 26). The vascular supply to the spathulate appendages departs from the common traces to the antesepalous stamens and antepetalous staminodes in a haphazard manner (Figure 27). The bundles branch repeatedly and supply the individual appendages (Figure 28). Our observations correspond more or less to those of Matthews et al. (2012). The arrangement of petal and stamen bundles in one plane is unusual but may be linked to the strong contortion of the flower in bud, when vasculature develops as a network of bundles in close vicinity of their respective organs.

DISCUSSION

Two Kinds of Staminodes in Ochnaceae?

The floral developmental evidence shows that *Sauvagesia* has pentacyclic flowers with two whorls of stamens and a tricarpellate gynoecium. The antepetalous stamen whorl is sterile and has become transformed into showy petaloid organs arranged as a cone around the stamens. Throughout eudicots there is a common tendency for one or more stamens or entire whorls to become reduced and either be lost, except for the stubs of the vascular traces, or develop new functions within the flower, i.e. staminodes (Walker-Larsen & Harder, 2000; Ronse De Craene and Smets, 2001). These staminodes frequently change form, becoming petal-like, i.e. petaloid, or can develop into nectaries, occasionally both.

Flowers in Sauvagesioideae are complex due to the presence of inner staminodes as well as outer appendages of the androecium arranged in one to several series. There is generally variation among genera between the presence of the two kinds of showy structures (e.g. *Sauvagesia* pro parte, *Neckia, Poecilandra*), the presence of only the inner petaloid structures (e.g. *Sauvagesia* pro parte, *Adenarake, Tyleria*), or only the outer filamentous structures (e.g. *Wallacea, Blastemanthus, Schuurmansia*) (Eichler, 1871; Gilg, 1893; Amaral, 1991). The outer spathulate or filiform appendages vary strongly in number, ranging from densely fasciculate to being completely suppressed and can be variable even within a same species (e.g. *Sauvagesia glandulosa, Sauvagesia* syn. *Lavradia montana*: Dwyer, 1945, 1964; Matthews et al. 2012).

There is some confusion in the distinction of these two kinds of "staminodes", although their presence is used as important characters for classification. Matthews et al. (2012)

distinguished six different types of staminodes in Ochnaceae: (1) numerous filamentous staminodes (*Blastemanthus*), (2) five small spathulate staminodes (some *Sauvagesia*), (3) an outer ring with broadened apex and an inner ring of filamentous staminodes (*Poecilandra*), (4) an outer ring with broadened apex and an inner whorl of petaloid staminodes (e.g. *Neckia*, *Sauvagesia*, this study), (5) two whorls of filamentous staminodes with the inner forming a tube (*Adenarake*), and (6) fused staminodes on the adaxial side of the flower by reduction of all stamens except one (*Testulea*). However, this distinction in six types is confusing as it does not clearly distinguish between the two kinds of appendages that we highlighted in *Sauvagesia erecta*. Dwyer (1945) considered both the interior petaloid appendages and the external filaments as staminodes, and described them respectively as interior and exterior corona. He also compared the occasional merging of the spathulate appendages to the fusion of the interior staminodes of *Lavradia* and the genus *Tyleria*.

Evidence for a Receptacular Origin of the Spathulate Appendages

Following arguments can be used in the discussion of the spathulate appendages as a corona of receptacular origin:

(1) In all cases for which developmental evidence is available, the corona appears very late in ontogeny, when all floral organs have been initiated and are already differentiated (e.g. Sajo et al., 2010; Waters et al., 2013; Bernhard, 1999; Amaral & Bittrich, 1998; this chapter). In the case of *Sauvagesia* the spathulate appendages arise after all other organs have been initiated and are without close connection with the androecium. Also in other genera (e.g. *Poecilandra, Blastemanthus*) outer filamentous appendages arise when anthers are differentiated (Amaral & Bittrich, 1998). For both the petaloid whorl and spathulate appendages to represent true staminodes, the antepetalous primordia would have to split to form two kinds of organs at an early developmental stage.

(2) Vasculature has been used as evidence for the staminodial or the perianth origin of the corona. A corona is usually supplied by vascular bundles, and this supply is often opportunistic, connecting the corona to the nearest available traces. Traces are sometimes inverted because of the late initiation of the corona (Puri, 1948; Sajo et al., 2010; Waters et al., 2013).

(3) The outer spathulate appendages of *Sauvagesia* are strongly reminiscent to the development of appendages in unrelated families, such as the external spines on the flower of *Neurada* (Neuradaceae) and *Agrimonia* (Rosaceae). The spines arise sequentially in alternation with the sepal lobes, but the development is highly comparable to the appendages of *Sauvagesia* (Ronse De Craene and Smets, 1996).

(4) The presence of two kinds of staminodes in a single flower is highly unusual in core eudicots. In Loasaceae staminodes produce variously elaborate nectar containers made up of generally five staminodes, which are part of a complex androecium containing fertile stamens (Hufford, 1990, 2003). Of these five staminodes three outer develop into a single cucullate structure, while the two inner elongate into various long appendages covering the stigma. However, in Loasaceae both types of

staminodes develop simultaneously with the rest of the androecium and are strongly connected to the fertile stamens.

(5) Diplostemony occurs in several Ochnoideae (e.g. *Campylospermum, Ouratea*), or the antepetalous whorl is sterile as in *Sauvagesia*. There are only a few cases of clearly haplostemonous flowers in the family, when an antepetalous staminodial whorl is missing (e.g. *Wallacea*). Obdiplostemony is absent, but the external position of antepetalous staminodes is linked with a delayed initiation as we could observe in *Sauvagesia* (cf. Amaral & Bittrich, 1998; Matthews et al., 2012). Several Ochnaceae, including Medusagynaceae and Quiinaceae are polyandrous, and Amaral (1991) suggested that polyandry might be plesiomorphic at the level of Ochnaceae s.str. The presence of numerous outer staminodes as in *Blastemanthus* would indicate a tendency for reduction to fewer stamens, which would be continued with the loss of the antepetalous stamens. However, in *Blastemanthus* there are two inner fertile stamen whorls corresponding to diplostemonous flowers in other genera, which makes a staminodial nature of the appendages suspicious. Although several other cases have been described as outer staminodial structures (see higher; Amaral, 1991; Matthews et al., 2012), the evidence for these as real staminodes needs to be verified by developmental studies.

Developmental studies in other Ochnaceae are currently limited (e.g. Amaral and Bittrich, 1998) but point to the very late appearance of the outer appendages. The very late appearance of the corona in *Sauvagesia* is linked with the development of an androgynophore and makes a receptacular origin more likely. The external appendages are best described as pseudostaminodes (cf. Ronse De Craene & Smets, 2001) or a corona. Here we confirm Eichler's (1871, 1878) assumption of a paracorollar structure, but not his interpretation of the petaloid staminodes as such.

The Significance of a Corona

The petaloid staminodial whorl of *Sauvagesia* is often described as a corona. The strict definition of a corona is that of a crown (latin origin). However, a corona is generally used indistinctly in a broad sense, sometimes referred to as a paracorolla (additional corolla), and various interpretations have been given about the homology and derivation of the corona from petals, tepals or stamens, including evidence from evo-devo (Hemingway et al., 2011; Waters et al., 2013). Ronse De Craene (2010) defines a corona as "a showy structure located between perianth and androecium of variable origin." This indicates that a corona is often not a proper organ but a novel structure that is not derived from a pre-existing one. A corona may arise as an appendage on the perianth (e.g. *Erythroxylon, Cardiospermum*), on the stamens (e.g. *Asclepias*), or on the receptacle (e.g. *Passiflora*). In other cases the corona takes its derivation directly from specific organs such as staminodes (e.g. *Hibbertia* sect. *pachynema* in Dilleniaceae, Loasaceae, *Sauvagesia*). When the corona takes excessive proportions, it can develop as a pseudocorolla, as is the case in *Sauvagesia*. The broad definition of the corona reflects a mixed bag of homologies, comparable to the definition of an epicalyx, which could be bract-derived or stipular (Ronse De Craene, 2010). For example, Gustafsson (2000) discussed the tubular corona of *Clusia gundlachii* as fused petals or staminodes. However, this cannot be verified ontogenetically as there is no evidence of fused petals and sympetaly is generally absent in Malpighiales, and other *Clusia* species with staminodial rings are different in several respects from *C. gundlachii*. The corona of *Narcissus* (Amaryllidaceae) has been interpreted as either androecial or tepalar in origin, although developmental evidence points to a hypanthial nature linked with a late development and epigyny (Waters et al., 2013). The corona of Passifloraceae (and probably Malesherbiaceae and Turneraceae) arises by extension of a hypanthium (Bernhard, 1999; Hemingway et al., 2011). The corona of some Lecythidaceae (e.g. Scytopetaloideae, *Napoleonaea, Asteranthos*) has been interpreted as staminodial in origin, but Ronse De Craene (2011) demonstrated that the so-called corona of *Napoleonaea* represents true petals developing in close relation to outer staminodial whorls.

Floral Adaptations for Pollination in *Sauvagesia*

The unusual development of an outer petaloid staminodial whorl enclosing stamens and carpels is shared by *Sauvagesia* and *Tyleria*. In *Tyleria* two additional outer appendages are attached outside the larger staminodes and may represent additional staminodial appendages (see figure 138 in Matthews et al. 2012), but this could only be confirmed by developmental studies. The tube is formed by a strong contortion of the staminodes, or by the fusion of staminodes into a cone (Matthews et al., 2012). This mechanism enhancing dispersion from poricidal anthers has been described as an example of transference of function (Kubitzki and Amaral, 1991). Buzz-pollination is generally associated with a syndrome of characters including poricidal anthers, reflexed petals, and an absence of nectaries (Renner, 1983; Proctor et al. 1996). No data are available on the pollination of *Sauvagesia*, although the staminodial cone acts as a pepper pot during buzz pollination funnelling the pollen very precisely from the dehiscing theca through the aperture (Figure 13) onto the abdomen of a visitor. The staminodial cone also has an important role in protecting the reproductive organs from pollen thieves. As many families of bees are involved in buzz pollination it is possible that damage to the stamens may occur inadvertently. There is however also evidence of large *Trigona* bees being pollen thieves and destroying poricidal stamens both in *Ochna* (Kubitzki & Amaral, 1991) and in Melastomataceae, (Renner, 1983). *Sauvagesia*, seems to have evolved a means of avoiding such damage by developing a protective tube of petaloid staminodes. Visually the resemblance of the staminodal cone to a large, colourful carpel may also help attract pollinators (Figure 14). This is enhanced by the contrast of the white petals and staminodial cone with the bright red corona. Given the contrast of the red appendages against the white petals and bicoloured petaloid staminodes it is highly probable that that they play a part in attracting pollinating insects, as they mimic an abundance of stamens, while the staminodal tube resembles an ovary. A similar mechanism of buzz pollination with folded petals has been reported in Elaeocarpaceae (Endress & Matthews, 2006).

CONCLUSION

Flowers of Ochnaceae have evolved a wide array of specialisations to assist with a buzz-pollination syndrome, leading to a high diversity in floral structures. The presumed presence of two kinds of staminodes in *Sauvagesia* is analysed with SEM and LM and it is concluded that the outer clusters of red spathulate appendages represent a corona of hypanthial origin, while the petaloid antepetalous structures represent real staminodes. The other genera of Sauvagesioideae are in need of an extended floral developmental study to understand the evolution of staminodes and filamentous appendages. This has been initiated by Amaral and Bittrich (1998), but awaits a more detailed investigation. We believe that diplostemony represents a plesiomorphic state for Ochnaceae and all Malphighiales. Understanding the process of a secondary stamen increase in polyandrous Ochnaceae, including Quiinaceae and Medusagynaceae will hopefully clarify this question.

From an evo-devo perspective the presence of petaloid staminodes is also fascinating as it shows a clear transference of petal characters to the antepetalous staminodes. This represents a case of homeosis, which deserves a more in-depth investigation of MADS box genes responsible for floral organ differentiation.

REFERENCES

Amaral, M. C. E. (1991). Phylogenetisch Systematik der Ochnaceae. *Botanische Jarhrbücher für Systematik, 113*, 105-196.

Amaral, M. C. E. & Bittrich, V. (1998). Ontogenia inicial do androceu de espécies de Ochnaceae subfam. Sauvagesioideae através da análise em microscopia eletrônica de varredura. *Revista Brasileira de Botânica, 21*, 269-273.

Angiosperm Phylogeny Group. [A. P. G.] (2009). An update of the Angiosperm Phylogeny Group classification for the orders and families of flowering plants, APG III. *Botanical Journal of the Linnean Society, 161*, 105-121.

Bernhard, A. (1999). Flower structure, development, and systematics in Passifloraceae and in *Abatia* (Flacourtiaceae). *International Journal of Plant Science, 160*, 135-150.

Cronquist, A. (1981). *An integrated system of classification of flowering plants.* New York, Columbia University Press.

Dwyer, J. D. (1945). The taxonomy of the genus *Sauvagesia* (Ochnaceae). *Bulletin of the Torrey Botanical Club, 72*, 521-540.

Dwyer, J. D. (1964). The taxonomy of *Lavradia* Vell. (Ochnaceae). *Bulletin du Jardin Botanique de L'Etat de Bruxelles, 34*, 507-518.

Eichler, A. W. (1871). Sauvagesiaceae. In, Martius, C. F. P. & Eichler, A. G., editors. *Flora brasiliensis*. Munich, Fleischer; 398-420.

Eichler, A. W. (1878). *Blüthendiagramme.*, vol. *2*. Leipzig, Engelmann.

Endress, P. K. & Matthews, M. L. (2006). Elaborate petals and staminodes in eudicots, diversity, function, and evolution. *Organisms, diversity and Evolution, 6*, 257-293.

Endress, P. K. & Matthews, M. L. (2012). Progress and problems in the assessment of flower morphology in higher-level systematics. *Plant Systematics and Evolution, 298*, 257-276.

Engler, A. (1876). Ochnaceae. In, Martius, C. F. P., Eichler, A. G. & Urban, I., editors. *Flora brasiliensis*. Munich, Fleischer; 297-366.

Gilg, E. (1893). Ochnaceae. In, Engler, A. & Prantl, K., editors. Die natürlichen Pflanzenfamilien III, 6. Leizig, Engelmann; 131-153.

Goebel, K. (1933). Organographie der Pflanzen insbesondere der Archegoniaten und Samenpflanzen Part 3. Ed. 3. Jena, Gustav Fischer.

Gustafsson, M. H. G. (2000). Floral morphology and relationships of *Clusia gundlachii* with a discussion of floral organ identity and diversity in the genus *Clusia*. *International Journal of Plant Science*, *161*, 43-53.

Hemingway, C. A., Christensen, A. R. & Malcomber, S. T. (2011). B- and C-class gene expression during corona development of the blue passionflower (*Passiflora caerulea*, Passifloraceae). *American Journal of Botany*, *98*, 923-934.

Hufford, L. D. (1990). Androecial development and the problem of monophyly of Loasaceae. *Canadian Journal of Botany*, *68*, 402-419.

Hufford, L. D. (2003). Homology and developmental transformation, models for the origins of the staminodes of Loasaceae subfamily Loasoideae. *International Journal of Plant Science*, *164* (5 Suppl.), S409-S439.

Iturralde, R. B. (2006). Notes on the Taxonomy and Distribution of the Ochnaceae in the Greater Antilles. *Willdenowia*, *36* (Special Issue), 455-461.

Kubitzki, K. & Amaral, M. C. E. (1991). Systematics and evolution transference of function in the pollination system of the Ochnaceae. *Plant Systematic Evolution*, *177*, 77-80

Matthews, M. L., Amaral, M. C. E. & Endress, P. K. (2012). Comparative floral structure and systematics in Ochnaceae *s.l.* (Ochnaceae, Quiinaceae and Medusagynaceae; Malpighiales). *Botanical Journal of the Linnean Society*, *170*, 299-392.

Puri, V. (1948). Studies in floral anatomy V. On the structure and nature of the corona in certain species of the Passifloraceae. *Journal of the Indian Botanical Society*, *27*, 130-149.

Renner, S. (1983). The widespread occurrence of anther destruction by Trigona bees in Melastomataceae. *Biotropica*, *15*, 251-256.

Proctor, M., Yeo, P. & Lack, A. (1996). *The Natural History of pollination*. Portland, Oregon, Timber Press.

Ronse De Craene, L. P. (2010). Floral diagrams. *An aid to understanding flower morphology and evolution*. Cambridge (UK), Cambridge University Press.

Ronse De Craene, L. P. (2011). Floral development of *Napoleonaea* (Lecythidaceae), a deceptively complex flower. In, Wanntorp, L. & Ronse De Craene, L. P. (eds), *Flowers on the tree of life*. Cambridge (UK), Cambridge University Press, 279-295.

Ronse Decraene, L. P. & Smets, E. (1996). The floral development of *Neurada procumbens* L. (Neuradaceae). *Acta Botanica Neerlandica*, *45*, 229-241.

Ronse De Craene, L. P. & Smets, E. F. (2001). Staminodes, their morphological and evolutionary significance. *Botanical Review*, *67*, 351-402.

Sajo, M. G. , de Mello-Silva R. & Rudall, P. J. (2010). Homologies of floral structures in Velloziaceae with particular reference to the corona. *International Journal of Plant Science*, *171*, 595-606.

Stevens, P. F. (2001 onwards). Angiosperm Phylogeny Website. Version 12, July 2012 (http://www. mobot. org/mobot/research/apweb/.

Walker-Larsen, J. & Harder, L. D. (2000). The evolution of staminodes in angiosperms, patterns of stamen reduction, loss, and functional re-invention. *American Journal of Botany*, *87*, 1367-1384.

Waters, M. T., Tiley, A. M. M., Kramer, E. M., Meerow, A. W., Langdale, J. A. & Scotland, R. W. (2013). The corona of the daffodil *Narcissus bulbocodium* shares stamen-like identity and is distinct from the orthodox floral whorls. *The Plant journal*, *74*, 615-625.

Wurdack, K. J. & Davis, C. C. (2009). Malpighiales Phylogenetics, Gaining Ground on one of the most Recalcitrant Clades in the Angiosperm Tree of Life. *American Journal of Botany*, *96*, 1551–1570.

Zappi, D. C. & Lucas, E. (2002) *Sauvagesia nitida* Zappi & E. Lucas (Ochnaceae), A New Species from Catolés, Bahia, NE Brazil, and Notes on *Sauvagesia* in Bahia & Minas Gerais. *Kew Bulletin*, *57*, 711-717.

ISBN: 978-1-62808-798-7
© 2013 Nova Science Publishers, Inc.

Chapter 6

POLLEN GRAIN DIAMETER: IN VITRO POLLEN GERMINATION AND REGRESSION BETWEEN GRAIN DIAMETER AND IN VITRO POLLEN GERMINATION IN PICKERELWEED (*PONTEDERIA CORDATA* L.)

Lyn A. Gettys

University of Florida IFAS, Department of Agronomy,
Fort Lauderdale Research and Education Center, Davie, FL, US

ABSTRACT

Pickerelweed (*Pontederia cordata* L.) is a tristylous species that utilizes heteromorphic incompatibility to reduce or prevent self-pollination. Three distinct floral morphs are produced by tristylous species, but each plant always produces flowers of the same morph. Previous reports have suggested that pollen produced by the three sets of anthers in pickerelweed differed from one another in grain diameter and in the length of pollen tubes generated during in vivo germination. A correlation has also been described between these variables, which suggests that pollen storage reserves play a role in compatibility of some combinations. The objective of this experiment was to verify previously reported grain diameter data and to determine whether in vitro pollen tube growth is influenced by grain diameter. Analysis of pollen from 12 plants (four each of S-morph, M-morph and L-morph) revealed that diameters of pollen grains produced by anthers borne by the three filament lengths of pickerelweed were significantly different from one another. Diameters of grains of s-pollen averaged 20.46 ± 0.34 μm, while mean diameters of m-pollen and l-pollen measured 35.04 ± 0.49 μm and 44.97 ± 0.34 μm, respectively. No overlap in grain diameter occurred among the three classes of pollen. Pollen tubes produced in vitro by l-pollen and m-pollen averaged 486.43 μm and 431.14 μm in length, respectively, 240 min after germination, while pollen tubes from s-pollen attained an average length of 265.57 μm. Previous reports suggested that pollen tube lengths produced in vivo by the three pollen diameter classes were significantly different from one another; however, I found no difference between lengths of pollen tubes from l-pollen and m-pollen produced during in vitro germination. The reason for these conflicting results is unknown but it is possible that other factors influence in vivo germination. A significant positive regression between pollen grain diameter and in vitro pollen tube length was identified; these results are similar to those described by other workers for in vivo pollen germination and suggest that pollen grain diameter has a positive influence on the length of pollen tubes produced during in vitro germination. This research provides evidence that pollen grain size and tube length may contribute to self-incompatibility in some, but not all, morph interactions in pickerelweed.

INTRODUCTION

Pickerelweed (*Pontederia cordata* L.) is a naturally outcrossed tristylous species that relies on heteromorphic incompatibility and herkogamy (the physical separation of reproductive structures) to reduce or prevent self-pollination. Some tristylous species are self-compatible, while others have degrees of self-incompatibility (Barrett 1988, 1993; Barrett and Anderson 1985; Darwin 1877; Eckert and Barrett 1994; O'Neill 1994). Three distinct floral morphs are produced by tristylous species, but each plant always produces flowers of the same morph. Floral morphs may differ from one another in characters including length or density of stigmatic papillae, style coloration and pollen exine sculpturing (Barrett 1988), but the most obvious visible difference among the floral morphs is style length. Stigmatic height variation is attributable to style length, as ovary length is similar in all three floral morphs (Richards and Barrett 1987).

There are three positions within each flower, with each position occupied by either a single style or one of two sets of stamens. Floral morph designation is determined by style length; flowers with long styles are L-morphs, while those with mid styles and short styles are classified as M-morphs and S-morphs, respectively (Figure 1).

Reciprocal positioning of anthers and stigmas occurs so that each plant produces flowers with anthers borne at the same level as the stigmas of the other floral morphs. This arrangement promotes insect-mediated cross-pollination between anthers and stigmas of equivalent height, resulting in seed set. Darwin (1877) referred to this as "legitimate pollination", while "illegitimate pollination" between anthers and stigmas at different levels results in little or no seed production.

Figure 1.The three floral morphs of pickerelweed. A) L-morph (long style). B) M morph (mid style). C) S morph (short style).

Figure 2.Anther arrangements in the three morphs of pickerelweed. A: Mid (m/L) and short (s/L) anthers of L-morph. B: Long (l/M) and short (s/M) anthers of M-morph. C: Long (l/S) and mid (m/S) anthers of S morph.

Several workers have described differences in the diameter of pollen grains produced by the three sets of anthers in pickerelweed. Three distinct pollen diameter classes are evident; anthers in the long position (borne by long filaments) produce pollen with the largest diameter and anthers in the short position (borne by short filaments) produce pollen with the smallest diameter. Anthers in the mid position (borne by mid-length filaments) produce pollen that is intermediate in diameter (Barrett and Glover 1985; Halsted 1889; Hazen 1918; Leggett 1875a, b; Ordnuff 1966; Price and Barrett 1982, 1984). Pollen produced by anthers borne on mid-length filaments is classified as m-pollen, while pollen produced by anthers borne on long or short filaments is classified as l-pollen or s-pollen, respectively. Pollen is further identified as s/M or s/L (s-pollen originating from M-morph plants or L-morph plants, respectively), m/S or m/L (m-pollen derived from S-morph plants or L-morph plants, respectively) and l/S or l/M (l-pollen from produced by S-morph plants or M-morph plants, respectively) (Figure 2).

Price and Barrett (1982, 1984) reported that there was no overlap in pollen diameter, so pollen origin (i.e., anther level) could be identified without ambiguity. Diameter classes are preserved regardless of whether pollen is fresh or acetolyzed (Barrett and Glover 1985; Price and Barrett 1984). Barrett and Glover (1985) reported that fresh grains of l-pollen measured 65.65 ± 3.22 μm in diameter, while fresh grains of m-pollen and s-pollen were 53.95 ± 3.60 μm and 34.52 ± 2.57 μm in diameter, respectively.

There are no differences among the floral morphs in regard to flower and inflorescence production, fecundity or ability to produce seed after cross-pollination, fruit weight or seed germination (Barrett and Anderson 1985; Price and Barrett 1982). The different floral morphs of pickerelweed exhibit varying levels of self-incompatibility, but all morphs produce more seeds after legitimate pollination than after illegitimate pollination (Barrett and Anderson 1985; Barrett and Glover 1985; Ordnuff 1966).

Anderson and Barrett (1986) found that pollen grains in both legitimate and illegitimate pollinations germinated readily on stigmas, which suggested that incompatibility in pickerelweed was not due to strong stigmatic inhibition. Anderson and Barrett (1986) showed that compatible pollen grains grew more quickly and reached the base of the style more often than incompatible pollen grains; exceptions were noted in pollinations of the S-morph, where all grains successfully reached the ovary. Legitimate pollination of all floral morphs resulted in pollen tubes that reached the base of the ovary within 2 h after pollination, while illegitimate pollinations took 4 h or longer to reach the base of the ovary. Pollinations of M-morphs and L-morphs with s-pollen rarely resulted in pollen tubes reaching the ovary and growth of the pollen tubes ceased after 8 h (Anderson and Barrett 1986).

Anderson and Barrett (1986) noted a correlation between pollen grain diameter and in vivo pollen tube growth. Pollen tubes from s-pollen, m-pollen and l-pollen reached 4 to 7 mm, 7 to 9 mm and 14 mm in length, respectively. Richards and Barrett (1987) found that stigmatic heights of S-morphs, M-morphs and L-morphs measured 2.7 ± 0.1 mm, 7.6 ± 0.3 mm and 12.6 ± 0.7 mm, respectively, with similar measurements recorded by Price and Barrett (1982). These results suggested that pollen storage reserves played a role in compatibility of some combinations, but Anderson and Barrett (1986) stated reduced seed set in illegitimate pollinations in spite of ovule penetration suggested the presence of an ovarian inhibitory system that may have retarded seed production in other combinations.

The objectives of this experiment were threefold. The first objective was to compare pollen grain diameter of same-level pollen produced by different floral morphs (s/M vs. s/L, m/S vs. m/L and l/S vs. l/M) and to identify differences in grain diameter among pollen produced by the three different anther levels (s-pollen vs. m-pollen vs. l-pollen). The second objective was to compare in vitro pollen tube growth of same-level pollen produced by different floral morphs (s/M vs. s/L, m/S vs. m/L and l/S vs. l/M) and to determine whether the differences in pollen tube growth in vivo described by Anderson and Barrett (1986) also exist in an in vitro system. The final objective of this experiment was to define the relationship between pollen grain diameter and in vitro pollen tube length 30, 60, 120 and 240 minutes after germination.

MATERIALS AND METHODS

Pollen grains from both anther levels of twelve plants were examined in this experiment; these comprised four each of L-morph plants (L1, L2, L3 and L4), M-morph plants (M1, M2, M3 and M4) and S-morph plants (S1, S2, S3 and S4). The plants used in this experiment were part of a population maintained for genetic and breeding studies at the University of Florida in Gainesville, FL, USA. Plants were grown in 1-L plastic nursery containers filled with a commercially available potting mix (MetroMix 510LL/MetroMix 500; Sun Gro Horticulture, Bellvue, WA,USA) that was amended with 10 g of controlled-release fertilizer (Osmocote Plus 15-9-12; The Scotts Co, Marysville, OH, USA) per container. Plants were sub-irrigated (water depth maintained at ca. 4 cm) and kept in a pollinator-free glasshouse with air temperature maintained at 27°C (day) and 16°C (night). Preliminary research revealed that some genotypes were more floriferous when grown under long days; therefore, supplemental lighting was used to artificially extend day length to 16 hours for the duration of this study.

Grains of l-pollen from eight plants were studied; four of these plants (S1, S2, S3 and S4) were S-morphs and the remaining four plants (M1, M2, M3 and M4) were M-morphs. Grains of m-pollen from eight plants were studied; four of these plants (L1, L2, L3 and L4) were L-morphs and the remaining 4 plants (S1, S2, S3 and S4) were S-morphs. Grains of s-pollen from eight plants were studied; four of these plants (M1, M2, M3 and M4) were M-morphs and the remaining four plants (L1, L2, L3 and L4) were L-morphs.

Pollen grain diameter. Dehisced anthers were removed from open flowers with fine forceps and placed in 2 mL of pollen killing and fixing solution in 6-well culture plates (BD Falcon™ Multiwell Cell Culture Plates #353046, BD Biosciences, Bedford MA; well volume 15 mL, well surface area 9.6 cm^2). The pollen killing and fixing solution consisted of 5 parts formaldehyde, 3 parts glacial acetic acid, 20 parts glycerin and 72 parts deionized water; this solution allowed hydration of the pollen grains but prevented germination. Three anthers bearing the same type of pollen from an individual plant were placed in each well. Each plate contained six wells, so an individual plate contained pollen from both anther levels of three different plants. Each plate assembly included a fitted cover, which was labeled with the source of the pollen in each well (donor identity and anther level).

Grain diameter was measured for 50 pollen grains from each plant/anther level combination. Grains were magnified and visualized using a Bausch and Lomb microprojector (Leica Microsystems, Wetzlar, Germany) equipped with a 10x eyepiece and a 20x objective for a final magnification of 395x. Diameters of magnified pollen grains were recorded in millimeters then converted to actual size in microns with a multiplier appropriate for the magnification used to visualize the sample. Means of converted values were calculated for each plant/anther level combination and these means were subjected to standard analysis of variance procedures (Steel et al. 1997). The model was constructed to identify differences among same-level pollen produced by plants with different floral morphs (l/S pollen vs. l/M pollen, m/S pollen vs. m/L pollen, s/M pollen vs. s/L pollen). The model also tested for differences among diameters of pollen grains produced by the three anther levels (l-pollen vs. m-pollen vs. s-pollen). Morph was nested within anther level, as each anther level was present in only two of the three morphs (l-pollen from S-morphs and M-morphs, m-pollen from S-morphs and L-morphs, s-pollen from M-morphs and L-morphs). Means were separated using t-tests to detect least significant differences (Steel et al. 1997).

In vitro pollen germination. This experiment utilized an agarose germination medium consisting of 10% sucrose, 0.6% agar, 0.02% Ca_3NO_4 and 0.01% boric acid dissolved in deionized water. This mixture was boiled for 5 minutes, then allowed to cool slightly before being transferred to the 6-well culture plates described above. Prepared culture plates were cooled to room temperature then stored at 4°C for up to 72 hours before pollen collection.

Pollen was collected from dehisced anthers at around 10:00 am. Fine forceps were used to remove anthers from open flowers and pollen was transferred to the surface of the germination medium by gently dragging the anthers across the surface of the medium. Each well was dusted with pollen from three anthers collected from the same level of an individual plant. Each plate contained six wells, so each plate contained pollen from both anther levels of three different plants. Each plate assembly included a fitted cover, which was labeled with the source of the pollen in each well (donor identity and anther level) and with the collection time. Plates were placed in a germination chamber maintained at 30°C and treated with killing and fixing solution at specified time intervals. Pollen grains from each anther level of each plant were killed at one of four time intervals: 30, 60, 120 and 240 minutes. All pollen samples in a single plate were killed at the same time interval by adding 2 mL of killing and fixing solution to each well.

Figure 3. Germinating pollen visualized utilizing a microprojector.

Pollen tube length was measured for 200 germinated pollen grains of each plant/anther level/interval combination. Pollen tubes were magnified and visualized at 395x using the Bausch and Lomb microprojector described above and tube length data were obtained by utilizing digital calipers to measure pollen tubes from the point of emergence from the pollen grain to the distal end of the pollen tube.

These data for magnified pollen tubes were recorded in millimeters, then converted to actual size in microns using a multiplier appropriate for the magnification used to visualize the sample.

Means of converted values were calculated for each plant/anther level/interval combination and these means were subjected to standard analysis of variance procedures (Steel et al. 1997). The model was constructed to identify differences in lengths of pollen tubes from same-level pollen produced by plants with different floral morphs (l/S pollen vs. l/M pollen, m/S pollen vs. m/L pollen, s/M pollen vs. s/L pollen) at all four intervals. The model also tested for differences among lengths of pollen tubes from pollen produced by the three anther levels (l-pollen vs. m-pollen vs. s-pollen) at all four intervals. Morph was nested within anther level, as each anther level was present in only two of the three morphs (l-pollen from S-morphs and M-morphs, m-pollen from S-morphs and L-morphs, s-pollen from M-morphs and L-morphs). Means were separated using t-tests to detect least significant differences (Steel et al. 1997).

Relationship between pollen grain diameter and tube length. The relationship between pollen grain diameter and tube length 30, 60, 120 and 240 minutes after germination was examined by computing a regression coefficient between the two variables. The Pearson product-moment correlation formula was used for this analysis (Steel et al. 1997).

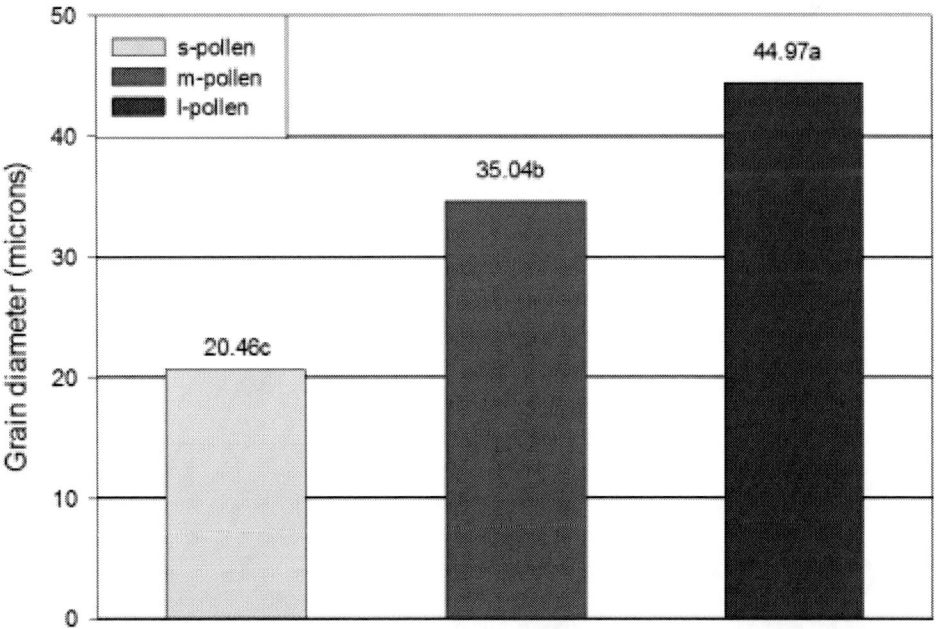

Figure 4. Grain diameter of pickerelweed. Bars represent the mean diameter of 400 grains per pollen class. Means were separated using a t-test to detect least significant differences. Grain diameters coded with different letters are significantly different at p=0.05.

CONCLUSION

Pollen grain diameter. There was no difference in same-level pollen grain diameter produced by the different floral morphs (p=0.2185), so further discussion of grain diameter will refer to pollen only by the anther level producing the pollen (i.e., l-pollen instead of l/S and l/M, m-pollen instead of m/S and m/L, s-pollen instead of s/M and s/L).

Significant differences were evident among grain diameters of l-pollen, m-pollen and s-pollen (p<0.0001). Grains of l-pollen averaged 44.97 ± 0.30 μm in diameter, while grains of m-pollen and s-pollen were 35.04 ± 0.49 μm and 20.46 ± 0.34 μm in diameter, respectively (Figure 4).

In vitro pollen germination. There was no difference in the lengths of pollen tubes generated by same-level pollen produced by the different floral morphs 30, 60, 120 or 240 minutes after germination (p=0.4650), so data for all pollen produced by the same anther level were pooled prior to comparisons among the three anther levels. Because floral morph did not have a significant impact on pollen tube length, further discussion of grain size will refer to pollen only by the anther level producing the pollen (i.e., l-pollen, m-pollen and s-pollen). Significant differences in pollen tube lengths were evident during in vitro germination of l-pollen, m-pollen and s-pollen (p<0.0001). Pollen tubes from l-pollen and m-pollen were longer than tubes from s-pollen at all time intervals under investigation.

Tubes from l-pollen and m-pollen were longer than tubes from s-pollen 30 minutes after germination, but there was no difference between tubes from l-pollen and tubes from m-pollen during the same time interval.

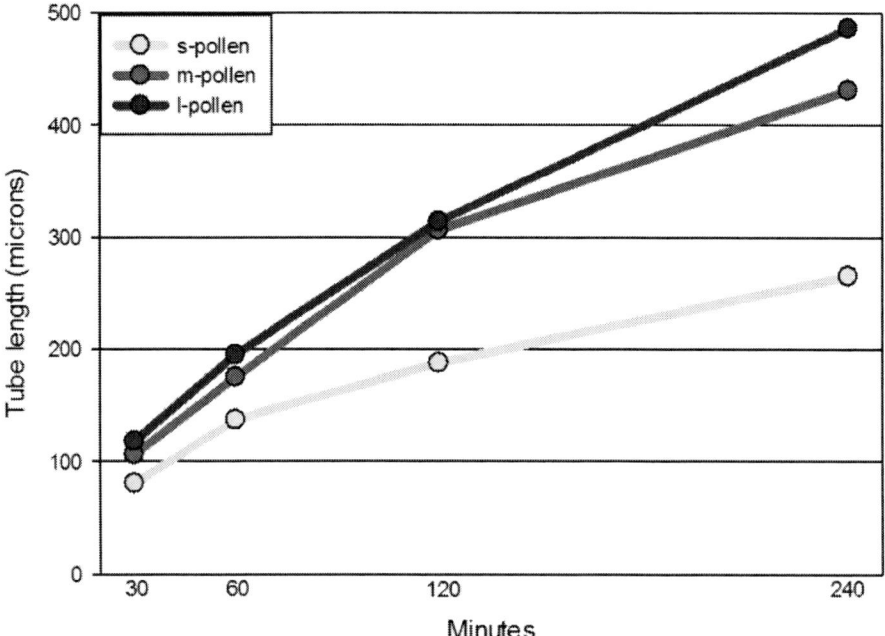

Figure 5. Mean length of pollen tubes in microns produced in vitro by pollen of pickerelweed 30, 60, 120 and 240 minutes after germination. Symbols represent the mean length of 2,000 pollen tubes for each anther level/interval combination.

Pollen tubes from l-pollen and m-pollen reached average lengths of 118.16 μm and 106.26 μm, respectively, 30 minutes after germination, while pollen tubes from s-pollen averaged 80.49 μm in length during the same interval (least significant difference 15.41 μm) (Figure 5).

Pollen tubes from l-pollen and m-pollen were longer than tubes from s-pollen 60 minutes after germination; in addition, pollen tubes from l-pollen were longer than pollen tubes from m-pollen during the same time interval.

Pollen tubes from l-pollen reached an average length of 195.07 μm 60 minutes after germination, while pollen tubes from m-pollen grew to 175.18 μm in length; s-pollen reached an average length of 137.27 μm during the same interval (least significant difference 17.34 μm) (Figure 5).

Pollen tubes from l-pollen and m-pollen were longer than pollen tubes from s-pollen 120 minutes after germination, but there was no difference between pollen tubes from l-pollen and pollen tubes from m-pollen during the same time interval. Pollen tubes from m-pollen and l-pollen reached average lengths of 306.34 μm and 313.74 μm, respectively, 120 minutes after germination, while pollen tubes from s-pollen averaged 187.77 μm in length during the same interval (least significant difference 63.62 μm) (Figure 5).

Pollen tubes from l-pollen and m-pollen were longer than pollen tubes from s-pollen 240 minutes after germination, but there was no difference between pollen tubes from l-pollen and pollen tubes from m-pollen during the same time interval.

Pollen tubes from l-pollen and m-pollen reached average lengths of 486.43 μm and 431.14 μm, respectively, 240 minutes after germination, while pollen tubes from s-pollen grew to an average length of 265.57 μm during the same interval (least significant difference 64.27 μm) (Figure 5).

Relationship between pollen grain diameter and pollen tube length. There was a highly significant regression between pollen grain diameter and pollen tube length at all time intervals investigated in this experiment. The result of this regression was an increase of 9.13 μm in in vitro pollen tube length for each micron increase in pollen grain diameter 240 minutes after germination, with similar trends noted at the three other intervals as well (Figure 6). These results reveal that pollen grain diameter has a significant positive impact on pollen tube growth in an in vitro system.

This experiment confirms the results of Barrett and Glover (1985) and Price and Barrett (1982, 1984) and shows that the diameters of pollen grains produced by the three anther levels of pickerelweed are significantly different from one another, with no overlap in grain diameter among the classes.

There was no difference in the diameter of same-level pollen grains from different floral morphs (e.g., l/S = l/M); therefore, pollen grain diameter is controlled by anther level or position (i.e., filament length).

Measurements of pollen grain diameters in this experiment differed from those reported by Barrett and Glover (1985) and Price and Barrett (1982, 1984); however, this was most likely a function of the disparate methods used to collect data as opposed to a true difference in grain diameter.

In vitro pollen germination did not produce the same results as those reported for in vivo germination by Anderson and Barrett (1986). Pollen tubes germinated in vitro did not generate the impressive growth reported in vivo, but this is not unexpected since pollen germination in vitro is often less robust than pollen germination in vivo.

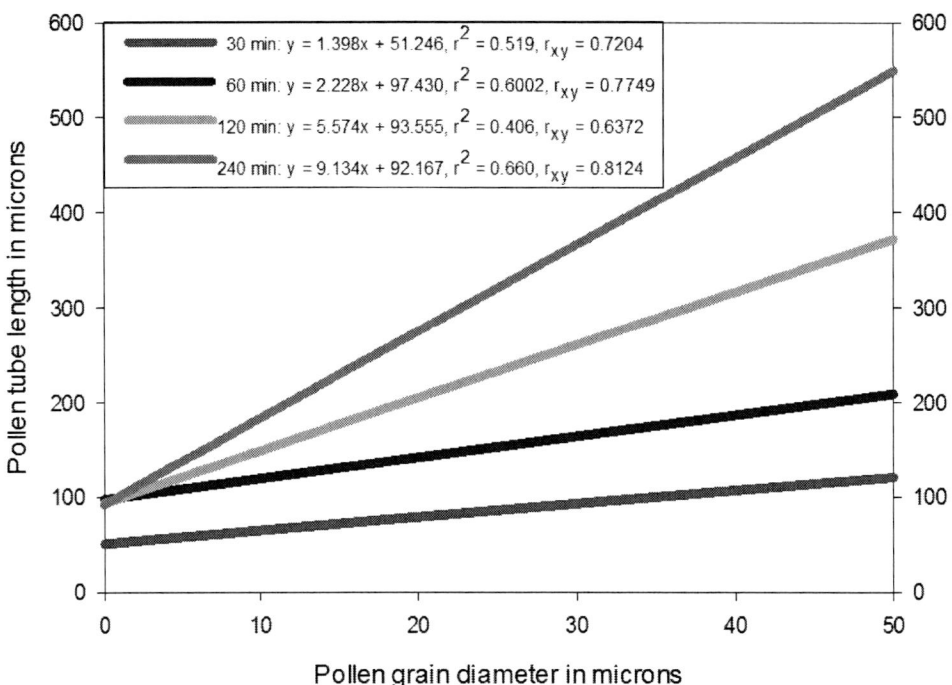

Figure 6.Regression between pollen grain diameter and in vitro tube length 30, 60, 120 and 240 minutes after germination. Regressions computed using 24 XY pairs with the mean diameter of 50 pollen grains (X) and the mean length of 200 pollen tubes (Y) for each plant/anther level/interval combination.

The relationships among pollen tubes from the three pollen grain diameter classes differed from those described by Anderson and Barrett (1986) as well; these workers reported significant differences among all three classes, but this experiment showed no significant difference between pollen tubes produced by l-pollen and those produced by m-pollen.

The reason for these conflicting results is unknown but it is possible that factors such as stylar interaction (e.g., the presence of inhibitory or stimulatory substances) with the germinating pollen grain influenced in vivo germination.

This experiment also detected a highly significant regression between pollen grain diameter and in vitro pollen tube length; these results are similar to those described by Anderson and Barrett (1986) for in vivo pollen germination and suggest that pollen diameter has a positive impact on the growth of pollen tubes produced as a result of in vitro germination.

These results also provide an explanation for Ordnuff's (1966) findings that self-pollinations of M-morphs and L-morphs were most fruitful when pollen from upper-level anthers (i.e., l/M × M and m/L × L) was used; as grain diameter conditions pollen tube length, larger pollen is more likely to produce a pollen tube long enough to travel down the length of the style and reach the ovary.

In addition, these findings support Anderson and Barrett's (1986) hypothesis that storage reserves play a role in compatibility of some combinations; however, Anderson and Barrett (1986) also pointed out that reduced seed set in illegitimate pollinations in spite of ovule penetration suggests the presence of an ovarian inhibitory system (i.e., somatoplastic incompatibility) that may retard seed production after self-pollination.

ACKNOWLEDGMENT

This research was supported by the University of Florida Institute for Food and Agricultural Sciences and by the Florida Agricultural Experiment Station. Mention of a trademark or a proprietary product does not constitute a guarantee or warranty of the product and does not imply its approval to the exclusion of other products that may be suitable.

I thank Wagner Vendrame for his thoughtful and constructive review of this manuscript. I also thank David Wofford, Paul Pfahler, David Sutton, Ramon Littell, Jody Strobel, Nancy Gaynor, Eastonce Gwata, and Eric Ostmark for their contributions to this experiment.

Reviewed by Wagner Vendrame, University of Florida IFAS, Tropical REC, Homestead, FL USA

REFERENCES

Anderson, J. M.; Barrett, S. C. H. 1986. Pollen tube growth in tristylous *Pontederia cordata* (Pontederiaceae). *Can. J. Bot.*,vol. 64, 2602-2607.

Barrett, S. C. H. 1988. Evolution of breeding systems in *Eichhornia* (Pontederiaceae): a review. *Ann. Missouri Bot. Garden*, vol. 75, 741-760.

Barrett, S. C. H. 1993. The evolutionary biology of tristyly. *In Oxford Surveys Evol. Biol.;* Futuyama, D.;Antonovics J.; Eds.; Oxford University Press, Oxford, UK, vol. 9, pp283-326.

Barrett, S. C. H.; Anderson, J. M. 1985. Variation of expression of trimorphic incompatibility in *Pontederia cordata* L. (Pontederiaceae). *Theoret. Appl. Genet.*, vol. 70, 355-362.

Barrett, S. C. H.; Glover, D. E. 1985. On the Darwinian hypothesis of the adaptive significance of tristyly. *Evolution,* vol. 39, 766-774.

Darwin, C. 1877. The different forms of flowers on plants of the same species. John Murray, London.

Eckert, C. G.; Barrett, S. C. H. 1994. Tristyly, self-compatibility and floral variation in *Decodon verticillatus* (Lythraceae) *Biol. J. Linn. Soc.*, vol. 53, 1-30.

Halsted, B. 1889. Pickerel weed pollen. Bull. *Torrey Bot. Club*,vol. 14, 255-257.

Hazen, T. 1918. The trimorphism and insect visitors of *Pontederia. Memoirs Torrey Bot. Club,* vol. 17, 459-484.

Leggett, W. H. 1875a. *Pontederia cordata* L. *Bull. Torrey Bot. Club*,vol. 6, 62-63.

Leggett, W. H. 1875b. *Pontederia cordata* L. *Bull. Torrey Bot. Club*, vol. 6, 170-171.

O'Neill, P. 1994. Genetic incompatibility and offspring quality in the tristylous plant *Lythrum salicaria* (Lythraceae). *Am. J. Bot.*, vol. 81, 76-84.

Ordnuff, R. 1966. The breeding system of *Pontederia cordata* L. *Bull. Torrey Bot. Club*, vol. 93, 407-416.

Price, S. D.; Barrett, S. C. H. 1982. Tristyly in *Pontederia cordata* L. (Pontederiaceae). *Can. J. Bot.*, vol. 60, 897-905.

Price, S. D.; Barrett, S. C. H. 1984. The function and adaptive significance of tristyly in *Pontederia cordata* L. (Pontederiaceae). *Biol. J. Linn. Soc.*,vol. 21, 315-329.

Richards, J. H.; Barrett, S. C. H. 1987. Development of tristyly in *Pontederia cordata* (Pontederiaceae). I. Mature floral structure and patterns of relative growth of reproductive organs. *Am. J. Bot.,* vol.74, 1831-1841.

Steel, R. G. D.; Torrie, J. H.; Dickey, D. A. 1997. Principles and procedures of statistics: a biometrical approach, 3rd ed. W.C. B. McGraw-Hill: New York, NY.

In: Flowers   ISBN: 978-1-62808-798-7
Editors: Teodor Berntsen and Kaj Alsvik   © 2013 Nova Science Publishers, Inc.

Chapter 7

DEVELOPMENT OF NOVEL POLLINATION TECHNIQUES TO OVERCOME THE EFFECTS OF HETEROMORPHIC INCOMPATIBILITY AND HERKOGAMY IN PICKERELWEED

Lyn A. Gettys

University of Florida IFAS, Department of Agronomy, Fort Lauderdale Research and Education Center, Davie FL, US

ABSTRACT

Pickerelweed is a primarily outcrossing tristylous species that utilizes herkogamy to reduce the likelihood of self-pollination. In addition to this physical separation of reproductive organs, pollen grains borne at the three positions within tristylous flowers produce tubes that correspond in length to reciprocally positioned styles. Most genetic studies require that plants be self-pollinated in order to determine gene action controlling traits and inbreeding can also be useful for developing inbred lines that may have utility in cultivar creation programs. The three morphs of pickerelweed have varying levels of self-incompatibility; M-morphs are self-fertile, whereas S- and L-morphs exhibit reduced fertility after normal self-pollination. In this study I developed techniques to reduce self-incompatibility in the S- and L-morphs of pickerelweed. Seed set in S-morphs self-pollinated using normal pollen transfer averaged 19.7%, whereas corolla removal increased seed production in these plants to an average of 43.6%. Seed set in L-morphs that were self-pollinated using normal methods averaged 2.7%, whereas stylar surgery of these plants increased average seed production to 29.4%. The methods developed in these experiments will be helpful for plant breeders and geneticists interested in studying this and other tristylous species.

INTRODUCTION

Genetic studies often require that the species under investigation be self-pollinated in order to determine the type of gene action and mode of inheritance of the trait of interest.

Pickerelweed (*Pontederia cordata* L.) is a primarily outcrossing tristylous species that employs heteromorphic incompatibility and herkogamy to reduce or prevent self-pollination. Members of tristylous species consistently bear one of three distinct floral types or morphs. Tristylous flowers house reproductive organs in three distinct positions; within each flower is a single style and two sets of anthers. Morph designation is determined by style length, i.e., long-styled flowers are referred to as L-morphs, while those with mid-length and short-length styles are classified as M-morphs and S-morphs, respectively. Anthers and stigmas are positioned in a reciprocal manner so that each plant produces flowers with anthers held at the same level as the stigmas of opposing morphs. A number of workers, including Darwin [1877] and Ordnuff [1966], have noted that this arrangement promotes pollinations between anthers and stigmas of equivalent height, which facilitates outcrossing and results in increased seed set in vivo as compared to self-pollination.

Anderson and Barrett [1986] reported that self-produced and foreign pollen grains both germinated readily on stigmas of pickerelweed, suggesting that self-incompatibility in the species is not due to strong stigmatic inhibition, but noted that few pollinations of M-morphs and L-morphs with s-pollen were successful because the pollen tubes rarely reached the ovary. These workers also identified a correlation between pollen grain size and in vivo pollen tube growth, which was confirmed by in vitro studies by Gettys [2013]. Pollen tubes generated during in vivo germination of s-pollen, m-pollen, and l-pollen reached 4-7 mm, 7-9 mm, and 14 mm, respectively [Anderson and Barrett 1986], whereas stigmatic heights of S-morphs, M-morphs, and L-morphs measure 2.7 ± 0.1 mm, 7.6 ± 0.3 mm, and 12.6 ± 0.7 mm, respectively [Price and Barrett 1982; Richards and Barrett 1987], which suggests that pollen storage reserves could play a role in seed set in some combinations.

The goals of this experiment were twofold. The first goal was to assess the level of self-incompatibility among members of a greenhouse population of pickerelweed and to assign a designation of self-compatible or self-incompatible to each member of the population. The second goal was to develop methods to overcome or bypass self-incompatibility mechanisms and to improve seed set after self-pollination in members of the population classified as self-incompatible.

MATERIALS AND METHODS

Plants used in this experiment were part of an experimental greenhouse population of pickerelweed created and maintained at the University of Florida in Gainesville. Plants were grown in 1-L nursery containers filled with a commercially available potting mix that was amended with 10 g of controlled-release fertilizer per container. Plants were sub-irrigated and kept in a pollinator-free glasshouse; daylength was maintained at 16 hours using supplemental lighting to ensure flower production and air temperature maintained at 27 °C day and 16 °C night.

All plants were pollinated using self-produced pollen from anther levels described by Ordnuff [1966] as being most productive (i.e., L-morphs were pollinated with m/L pollen, M-morphs with l/M pollen, and S-morphs with m/S pollen). Anthers were removed from flowers with fine forceps, and pollen transfer was accomplished by brushing the stigma with an anther

(L- and M-morph flowers) or by depositing a whole anther deep in the throat of the flower (S-morph flowers). Magnifying headgear was worn during all pollinations to allow visual confirmation of successful transfer and adhesion of pollen grains to the stigma. Forceps were flame-sterilized between pollination of different plants to prevent contamination with foreign pollen. Pollinations commenced with the opening of the first flowers of an inflorescence and continued until all flowers on the inflorescence had been pollinated (between 7 and 12 days). All pollinations were performed between 10 am and 2 pm daily, and all flowers in each inflorescence were pollinated using the same method. Daily pollination data were recorded on jewelry tags placed on each inflorescence. Each completed inflorescence was enclosed in a small tulle bag and secured with a plastic-covered twist-tie until fruits were ripe. Fruits were considered ripe when the bearing infructescence shattered (usually 23 to 30 days after completion of pollinations).

Fruits were collected in their tulle bag and air-dried for 7 days, then de-husked using a rubber-covered rub board. The use of the rub board allowed removal of the outer husk of the fruit without scarification of the enclosed seed. Percent seed set was calculated by dividing total number of seeds by total number of flowers pollinated. Genotypes that produced less than 10% seed set using the methods described above were classified as self-incompatible and subjected to one of two novel pollination techniques. All M-morph genotypes were self-compatible; therefore, development of a novel pollination technique for M-morph flowers was unnecessary.

L-morph genotypes deemed self-incompatible were treated with stylar surgery. The thumbnail of one hand was placed under the style of the flower; the style was then shortened with a surgical blade held between the thumb and forefinger of the other hand. The style was cut so that the tip of the style was ca. the same length as the mid-level anthers, and pollen was immediately transferred from the mid-level anthers to the cut tip of the style. All flowers on each inflorescence were pollinated using stylar surgery, and completed inflorescences were treated as described above.

The floral envelope was completely removed from self-incompatible S-morph genotypes to increase access to the stigma. Removal of the floral envelope was accomplished by firmly grasping the center of the flower with forceps and gently pulling to remove the corolla. This protocol effectively exposed the stigma in most cases; however, some plants required further manipulation to remove tepal tissue obstructing the stigma. Removal of the floral envelope resulted in emasculation of the flower, so mid-level pollen was taken from the removed portion of the flower and transferred to the exposed stigma. All flowers on each inflorescence were pollinated in this manner, and completed inflorescences were treated as described above. Multiple inflorescences were pollinated on most plants to ensure production of sufficient quantities of seeds.

Percentage data were normalized using a modified arcsine transformation as described by Zar (1996); this equation is:

$$p' = 1/2 \left[\arcsin \sqrt{(^X/_{n+1})} + \arcsin \sqrt{(^{X+1}/_{n+1})} \right]$$

where p′ represents the transformed percentage, p symbolizes the original percentage data , and X/n symbolizes the actual proportion data.

CONCLUSION

Mid Morph Plants

Seed set in "legitimate" cross-pollinations of M-morph plants ranged from 78.9% to 94.0% (mean 88.5%) (data not shown). A total of 14,425 flowers were self-pollinated on 53 M-morph plants (mean flowers pollinated per plant = 291.04). Seed set ranged from 37.3% to 98.8% (mean 73.65%), with 10,979 seeds produced (mean seeds produced per plant = 207.15). All M-morph genotypes studied in this experiment were thus classified as self-compatible. This high level of seed production after self-pollination corresponds well with previous reports by others, including Barrett and Anderson [1985], Barrett and Glover [1985] and Ordnuff [1966], that self-incompatibility is weakest in M-morph flowers of pickerelweed.

Long Morph Plants

Seed set in "legitimate" cross-pollinations of L-morph plants ranged from 61.3% to 83.1% (mean 77.7%) (data not shown). 51 L-morph plants were self-pollinated in this experiment, with 28 of these plants classified as self-compatible, with seed set around half that of cross-pollinated plants. Self-pollination of 13,412 flowers in the self-compatible group (mean 479 flowers per plant) produced a total of 4,429 seeds (mean 158.18 seeds per plant). Seed set ranged from 15.3% to 66.3% (mean 35.7%). As these plants are self-compatible, stylar surgery was not performed on them and further data regarding this group will not be presented. The 23 remaining L-morph plants were considered self-incompatible. Normal (control) self-pollination of 9,237 flowers in this group (mean 401.61 flowers per plant) resulted in the production of only 262 seeds (mean 11.39 seeds per plant) (Figure 1). Seed set resulting from normal self-pollination of these plants ranged from 0 to 9.8%, with a mean seed set of 2.7%. Stylar surgery and subsequent self-pollination of 11,049 flowers (mean 480.39 flowers per plant) in the self-incompatible group resulted in the production of 3,248 seeds (mean 141.22 seeds per plant), with seed set ranging from 14.7% to 61.6% (mean 29.4%) (Figure 1).

Data from L-morph plants subjected to stylar surgery were transformed as described above and F-tests were conducted to determine whether homogeneity of variances existed between the control pollinations and the treated (stylar surgery) pollinations. Arcsine transformation of the data resulted in homogeneity of variances. Transformed data were then subjected to an unpaired t-test, which revealed significant differences between the means of the control and stylar surgery datasets. Strong evidence for differences between the two pollination datasets was also presented when a t-test was performed on a dataset composed of the paired differences between control and stylar surgery pollinations. Gonick and Smith [1993] suggest that paired comparisons are useful, as the analysis eliminates variability among plants and experimental variance is therefore limited to intra-plant variability. These results suggest that the stylar surgery treatment is an effective method to improve seed set in L-morph plants that have been classified as self-incompatible when pollinated using normal techniques.

Plant no.	Control (normal)		Stylar surgery	
	Seed set/ pollinations	% seed set	Seed set/ pollinations	% seed set
L1	0/150	0.0	146/482	30.3
L2	1/272	0.4	117/479	24.0
L3	3/280	1.1	111/604	17.3
L4	35/522	6.1	85/369	16.9
L5	6/500	1.2	98/457	20.2
L6	1/313	0.3	124/565	21.6
L7	0/47	0.0	76/518	14.7
L8	19/437	4.3	69/364	19.0
L9	3/476	0.6	354/875	39.9
L10	26/372	7.0	156/412	30.9
L11	30/958	3.1	176/527	30.3
L12	0/385	0.0	137/512	26.8
L13	0/501	0.0	189/713	26.5
L14	8/408	2.0	129/488	23.4
L15	17/230	7.4	149/242	54.2
L16	22/421	5.2	148/446	28.0
L17	2/238	0.8	131/409	31.2
L18	2/489	0.4	101/452	21.9
L19	1/537	0.2	119/465	25.4
L20	50/512	9.8	201/471	32.9
L21	19/319	6.0	111/226	43.1
L22	1/360	0.3	148/439	33.4
L23	19/510	3.7	173/534	28.7

Figure 1. Seed set of L-morph plants with and without stylar surgery.

It is likely that the pollen tube growth limitations described by Anderson and Barrett [1986] are at least partly responsible for self-incompatibility and poor seed production after normal self-pollination of the L-morph of pickerelweed. The largest pollen grain produced by the L-morph is m-pollen; Anderson and Barrett [1986] found that m-pollen forms a pollen tube that is 7-9 mm in length, but Price and Barrett [1982] and Richards and Barrett [1987] found that stigmas of L-morphs measure 12.6 ± 0.7 mm in length. The stylar surgery employed in this experiment artificially shortened the style and therefore the travel distance required for a pollen tube to reach the ovule and effect fertilization.

Roggen and van Dijk [1972] used a steel brush to simultaneously mutilate and pollinate the stigma of *Brassica oleracea* L. in order to increase seed set, but this species is self-incompatible due to sporophytic factors (i.e., the stigma is the site of inhibition), so the goal of their experiment was to eliminate the stigmatic barriers responsible for incompatibility. Stylar surgery is much more similar to the stump pollination technique employed by Davies [1957] to facilitate seed set in interspecific crosses between members of the genus *Lathyrus*. The two species investigated by Davies, *L. odoratus* and *L. hirsutus*, produce styles of different lengths; the length of the style of *L. odoratus* is 10 mm, while the length of the style of *L. hirsutus* is 4 mm. The interspecific cross-pollination of *L. hirsutus* x *L. odoratus* results in fertilization, but the reciprocal event normally fails to produce seeds. Davies [1957] removed the stigma and part of the style of *L. odoratus* and applied pollen from *L. hirsutus* to the cut end of the style; this resulted in fertilization as the pollen had to travel a shorter distance to reach the ovary.

Short Morph Plants

Seed set in "legitimate" cross-pollinations of S-morph plants ranged from 14.5% to 56.5% (mean 41.8%) (data not shown). 38 S-morph plants were pollinated using both the normal (control) and the corolla removal techniques. Control self-pollination of 10,805 flowers in this group (mean 284.3 flowers per plant) resulted in the production of 2,100 seeds (mean 55.3 seeds per plant). Seed set resulting from normal self-pollination of these plants ranged from 4.6% to 42.9%, with a mean seed set of 19.7% (Figure 2). Corolla removal and subsequent self-pollination of 11,086 flowers (mean 297.1 flowers per plant) of the same plants resulted in the production of 4,839 seeds (mean 127.3 seeds per plant), with seed set ranging from 18.0% to 82.3% (mean 43.6%) (Figure 2). Removal of the floral envelope was a reasonably simple manipulation and early results showed that seed set was greatly increased following removal of the floral envelope; therefore, all S-morph plants pollinated during the last few months of this experiment were subjected to this protocol. A group of 20 plants were pollinated using only the corolla removal treatment. A total of 3,380 seeds (mean 169.0 seeds per plant) were produced after corolla removal pollination of 7,627 flowers (mean 381.25 flowers per plant); seed set ranged from 13.6% to 78.1% and averaged 44.3%. Since these plants were only pollinated using the corolla removal technique, further data relating to this group will not be described.

Data from S-morph plants subjected to both pollination protocols were transformed as described above and F-tests were conducted to determine whether homogeneity of variances existed between the control pollinations and the treated (corolla removal) pollinations. Arcsine transformation of the data resulted in homogeneity of variances. Transformed data were then subjected to an unpaired t-test, which showed that the means of the control and corolla removal pollination datasets were different. Further evidence for differences between the two datasets was provided when a t-test was performed on a dataset composed of the paired differences between control and corolla removal populations. These results suggest that the corolla removal technique is an effective method to improve seed set in S-morph plants when compared to the same plants pollinated using normal techniques.

It is likely that reduced access to the stigma is at least partly responsible for self-incompatibility and poor seed production after normal self-pollination of the S-morph of pickerelweed. Price and Barrett [1982] and Richards and Barrett [1987] stated that stigmas of S-morph flowers measure only 2.7 ± 0.1 mm in height, and gross visual observation reveals that the reproductive structure is ensconced deep within the throat of the flower. Barrett and Anderson [1985] used forceps to split the floral perianth of S-morph flowers to increase access to the stigma and anthers, but stated that it was still difficult to conduct pollinations using S-morph flowers as seed parents. This study found that removal of the floral envelope to allow increased access to the stigma was a reasonably simple manipulation, and the resulting increased seed set was well worth the small effort required.

Self-incompatibility and the resultant poor seed set in some floral morphs of pickerelweed may be overcome with the use of the novel pollination techniques developed, tested, and described in this experiment. Some workers [i.e., Anderson and Barrett 1986] have stated that poor seed set after self-pollination may be due to the presence of an ovarian inhibitory system that retards seed production after self-pollination. This study suggests that physical constraints (e.g., style length in L-morphs and stigma access in S-morphs) play an important role in the prevention of self-pollination in pickerelweed, and can be bypassed to effect adequate production of seeds after self-pollination. This information will be helpful for plant breeders and geneticists interested in studying this and other tristylous species. Geneticists can use this information to improve seed set after self-pollination, as most genetic studies require at least one generation of inbreeding to determine the mode of inheritance and gene action of the trait under investigation. Plant breeders may employ these techniques to develop inbred lines of tristylous species, barring the presence of severe inbreeding depression in the species of interest.

Plant no.	Control (normal)		Corolla removal	
	Seed set/ pollinations	% seed set	Seed set/ pollinations	% seed set
S1	24/111	21.6	110/325	33.8
S2	40/264	15.2	252/486	52.0
S3	10/53	18.9	161/444	36.3
S4	36/720	16.4	113/467	24.2
S5	75/428	17.5	61/230	26.5
S6	59/384	15.4	116/250	46.4
S7	35/130	26.9	148/262	56.5
S8	106/286	37.1	54/118	45.8
S9	77/315	24.4	105.147	69.4
S10	19/96	19.8	117/332	53.3
S11	70/186	37.6	161/228	70.6
S12	67/185	36.5	47/100	47.0
S13	157/366	42.9	99/160	61.9
S14	61/418	14.6	131/353	37.1
S15	30/333	9.0	92/192	47.9
S16	78/374	20.9	173.270	64.1
S17	19/183	10.4	197/426	46.2
S18	71/558	12.7	86/188	45.7
S19	100/572	17.5	65/224	29.0
S20	66/169	39.1	56/109	51.4
S21	117/625	18.7	64/211	30.3
S22	48/189	25.4	334/473	70.6
S23	36/372	9.7	92/512	18.0
S24	63/329	19.1	127/272	46.7
S25	11/99	11.1	134/247	54.3
S26	35/325	10.8	65/287	22.6
S27	12/124	10.5	160/326	49.1
S28	43/308	14.0	114/396	28.8
S29	34/168	20.2	173/417	41.5
S30	5/85	5.9	143/417	34.3
S31	19/149	12.8	243/551	44.1
S32	83/408	20.3	117/263	44.5
S33	17/371	4.6	131/451	29.0
S34	84/385	21.8	92/155	59.4
S35	74/412	18.0	66/133	49.6
S36	96/322	29.8	87/149	58.4
S37	80/356	22.5	96/200	48.0
S38	42/117	35.9	260/316	82.3

Figure 2. Seed set of S-morph plants with and without corolla removal.

ACKNOWLEDGMENTS

This research was supported by the University of Florida Institute for Food and Agricultural Sciences and by the Florida Agricultural Experiment Station. Mention of a trademark or a proprietary product does not constitute a guarantee or warranty of the product and does not imply its approval to the exclusion of other products that may be suitable. I would like to thank David Wofford, David Sutton, Paul Pfahler and Jody Strobel for their contributions to these experiments.

REFERENCES

Anderson, J. M. & Barrett, S. C. H. (1986). Pollen tube growth in tristylous *Pontederia cordata* (Pontederiaceae). *Can. J. Bot.*, vol. *64*, 2602-2607.

Barrett, S. C. H. & Anderson, J. M. (1985). Variation of expression of trimorphic incompatibility in *Pontederia cordata* L. (Pontederiaceae). *Theor. Appl. Genet.*, vol. *70*, 355-362.

Barrett, S. C. H. & Glover, D. E. (1985). On the Darwinian hypothesis of the adaptive significance of tristyly. *Evolution*, vol. *39*, 766-774.

Darwin, C. (1877). The different forms of flowers on plants of the same species. John Murray, London.

Davies, A. J. S. (1957). Successful crossing in the genus *Lathyrus* through stylar amputation. *Nature* (London), vol. *180*, 612.

Gettys, L. A. (2013). Pollen grain diameter, in vitro pollen germination and regression between grain diameter and in vitro pollen germination in pickerelweed (*Pontederia cordata* L.). In: Editor EE, editor. Flowers: Morphology, evolutionary diversification and implications for the environment. Nova Science Publishers, Hauppauge, NY, XX-XX.

Gonick, L. & Smith, W. (1993). The cartoon guide to statistics. HarperCollins, New York, NY.

Ordnuff, R. (1966). The breeding system of *Pontederia cordata* L. Bull. *Torrey Bot. Club*, vol. *93*, 407-416.

Price, S. D. & Barrett, S. C. H. (1982). Tristyly in *Pontederia cordata* L. (Pontederiaceae). *Can. J. Bot.*, vol. *60*, 897-905.

Richards, J. H. & Barrett, S. C. H. (1987). Development of tristyly in *Pontederia cordata* (Pontederiaceae). I. Mature floral structure and patterns of relative growth of reproductive organs. *Am. J. Bot.*, vol. *74*, 1831-1841.

Roggen, H. P. Jr. & van Dijk, A. J. (1972). Breaking incompatibility in *Brassica oleracea* L. by steel brush pollination. *Euphytica*, vol. *21*, 424-425.

Zar, J. H. (1996). Biostatistical analysis, 3rd ed. Prentice Hall, Upper Saddle River, NJ.

Reviewed by Kevin Kenworthy, University of Florida IFAS, Department of Agronomy, Gainesville FL USA.

INDEX

edible chrysanthemum, vii, 1, 5, 6, 7, 8, 17, 18, 30
editors, 101
EEG, 79
electromagnetic, 61
e-mail, 55
emission, viii, 55, 61, 64, 65
energy, 34
energy supply, 34
England, 12, 14, 79
enteritis, 80
environment(s), vii, 10, 34, 47, 49, 56, 62, 66, 124
enzyme(s), 77, 78, 78, 82
epidemiologic, 76
epidermis, 97
epithelial cells, 19
Epstein-Barr virus, 13
equipment, 47
erosion, viii, 33, 47
ester, 13, 15, 19, 20
esthetics, vii, 1
ethanol, 11
ethyl acetate, 82
ethyl alcohol, 24
Eurasia, 14, 18
Europe, 12, 16, 19, 24
European Union, 22
evidence, x, 24, 27, 65, 76, 77, 84, 91, 97, 98, 99, 100, 106, 120, 122
evil, 78
evolution, 51, 52, 70, 101, 102
excitation, 61
exclusion, 57, 115, 124
expulsion, 90
extraction, 72, 85
extracts, 4, 5, 7, 9, 14, 19, 22, 23, 28, 31, 72, 78, 79, 80, 82, 83, 84, 85, 86

F

fainting, 79
falciparum malaria, 14
families, 90, 98, 100, 101
fear(s), 77, 83
FEMA, 82
ferric ion, 65, 82
fertility, x, 46, 47, 117
fertilization, 49, 121
fibers, 50, 51, 53
fibroblasts, 84
field tests, 48
filament, ix, 105, 113
filiform, 91, 97
financial support, 69

fish oil, 28
fixation, 35, 70
flame, 119
flatulence, 80
flavonoids, vii, 1, 7, 15, 16, 17, 18, 21, 23, 30, 31, 55, 64, 82, 85
flavonol, 4, 13, 14, 20
flavor, viii, 55, 67, 68, 76, 78
flight, 64, 72
fluorescence, viii, 55, 64, 65, 71, 73
folic acid, 78
folklore, 79
food, 3, 4, 8, 22, 71, 72, 76, 77, 78, 81, 82
Food and Drug Administration, 82, 84
food industry, 81
food products, 76
formaldehyde, 109
formation, 12, 17, 92
formula, 111
free radicals, 77
frost, 48
fruits, 34, 35, 36, 69, 119
functional food, 28
fusion, 98, 100

G

gastritis, 80
gastroenteritis, 19
gel, 19
gene expression, 19, 102
genes, 101
genetic diversity, 42, 52
geneticists, x, 117, 123
genetics, 53
genome, 40, 41, 42, 43, 51, 52
genus, vii, 3, 12, 13, 14, 16, 18, 21, 33, 34, 35, 36, 41, 42, 43, 49, 50, 51, 52, 76, 85, 86, 90, 91, 98, 101, 102, 121, 124
geography, 43
Georgia, 44
Germany, 26, 32, 109
germination, vii, ix, 105, 108, 109, 110, 111, 112, 113, 114, 118, 124
ginger, 14
glucose, 10
glucoside, 4, 17, 20, 21, 23
glutamate, 79
glutathione, 86
glycerin, 109
glycol, 49
glycoside, 24
God, 82